TECHNICAL REPORT

Integrating the Department of Defense Supply Chain

Eric Peltz • Marc Robbins

With *Geoffrey McGovern*

Prepared for the Office of the Secretary of Defense

Approved for public release; distribution unlimited

The research described in this report was prepared for the Office of the Secretary of Defense (OSD). The research was conducted within the RAND National Defense Research Institute, a federally funded research and development center sponsored by OSD, the Joint Staff, the Unified Combatant Commands, the Navy, the Marine Corps, the defense agencies, and the defense Intelligence Community under Contract W74V8H-06-C-0002.

Library of Congress Cataloging-in-Publication Data

Peltz, Eric, 1968-
Integrating the Department of Defense supply chain / Eric Peltz, Marc Robbins, with Geoffrey McGovern.
p. cm.
Includes bibliographical references.
ISBN 978-0-8330-7641-0 (pbk. : alk. paper)
1. United States—Armed Forces—Supplies and stores. 2. Logistics—United States. 3. United States. Defense Logistics Agency. 4. United States. Dept. of Defense—Rules and practice—Evaluation. I. Robbins, Marc. II. McGovern, Geoffrey. III. United States. Deputy Under Secretary of Defense (Logistics and Materiel Readiness) IV. Title.

UC263.P455 2012
355.6'210973—dc23

2012033339

Published 2012 by the RAND Corporation
1776 Main Street, P.O. Box 2138, Santa Monica, CA 90407-2138
1200 South Hayes Street, Arlington, VA 22202-5050
4570 Fifth Avenue, Suite 600, Pittsburgh, PA 15213-2665
RAND URL: http://www.rand.org/
To order RAND documents or to obtain additional information, contact
Distribution Services: Telephone: (310) 451-7002;
Fax: (310) 451-6915; Email: order@rand.org

Preface

Recognizing the promise of achieving a more integrated supply chain, the Assistant Secretary of Defense for Logistics and Materiel Readiness (ASD(L&MR)) sponsored the project that led to this report. The project was intended to provide an informed perspective on how the Department of Defense (DoD) supply chain could become more integrated based upon the body of RAND Corporation research since the mid-1990s that has focused on improving DoD supply chain performance and efficiency and based upon recent and ongoing DoD supply chain management improvement initiatives. This report draws these threads together and is intended to provide a guide for the design and improvement of DoD supply chain policy, structure, and management practices. It should also be valuable for DoD supply chain personnel at all levels with respect to understanding and determining how to improve the role they play in maximizing overall supply chain cost-effectiveness.

The DoD sustainment supply chain community has increased performance and harvested significant efficiencies through process improvement activities and rationalization of common activities. However, the majority of strides have been made within functions and processes. We posit that more opportunities for improvement remain in end-to-end supply chain integration—spanning all DoD organizations and its suppliers—of processes that jointly affect total supply chain costs and performance.[1] The report explains what is meant by supply chain integration, provides illustrative evidence of DoD supply chain integration shortfalls, and describes why there are shortfalls in integration. It then provides a framework for an integrated DoD supply chain, associated recommendations for DoD supply chain policy, and a framework for developing management practices that drive people to take actions that produce supply chain integration. In the course of the project, the ASD(L&MR) adopted many of the policy recommendations in the drafting of an update of DoD supply chain materiel management policy; these changes are also described in the report. The report then turns to potential opportunities to improve DoD supply chain efficiency and performance built on the earlier material. These opportunities also provide further indications of the room to improve supply chain integration. The ASD(L&MR) and other DoD supply chain organizations have begun pursuing some of these, as indicated in the report.

This research was conducted by the Forces and Resources Policy Center of the RAND National Defense Research Institute (NDRI). NDRI is a federally funded research and development center sponsored by the Office of the Secretary of Defense, the Joint Staff, the Unified

[1] The scope of this report and the associated project is the DoD sustainment supply chain, with an emphasis on classes II (clothing, individual equipment, tools, and administrative supplies), IIIP (packaged petroleum, oil, and lubricants), IV (construction materiel), and IX (repair parts).

Combatant Commands, the Navy, the Marine Corps, the defense agencies, and the defense Intelligence Community.

For more information on the RAND Forces and Resources Policy Center, see http://www.rand.org/nsrd/ndri/centers/frp.html or contact the director (contact information is provided on the web page).

Contents

Figures

Tables

Summary

In the mid-1990s, the U.S. Department of Defense (DoD) began a sustained effort to improve its supply chain, improving performance and harvesting significant efficiencies through process improvement initiatives, rationalizing functional activities across organizations, and integrating functions and organizations within processes. However, additional opportunity exists for integrating the supply chain across processes. In a fully integrated supply chain, processes are intertwined in a way that process design and execution decisions must consider impacts on all other processes and the total supply chain in order to achieve optimal supply chain performance and efficiency rather than focusing on the success of individual processes, functions, and organizations.

To help DoD determine how to tap the full potential of supply chain integration, the Assistant Secretary of Defense for Logistics and Materiel Readiness (ASD(L&MR)) asked the RAND National Defense Research Institute (NDRI), based upon prior research and analysis and ongoing DoD initiatives, to develop a framework for an integrated DoD supply chain, identify barriers and enablers to integration, and make recommendations to align policy with the framework. In addition, NDRI was also asked to identify opportunities for efficiency through improved integration.[2]

Case Studies

The project developed two related case studies that illustrate the need for improvement in DoD supply chain integration.

- The first is on the DoD journey to improve centralized theater inventory, which focuses on optimizing the trade-offs among inventory, transportation, and materiel handing to minimize total supply chain costs versus focusing on minimizing each of these costs independently.
- The second case study shows how one functionally isolated decision—a well-meaning decision to shift transportation modes to reduce costs—propagated across the supply

[2] The study's scope included supply classes II (clothing, individual equipment, tools, and administrative supplies), IIIP (packaged petroleum, oil, and lubricants), IV (construction materiel), VI (personal demand items), and IX (repair parts). These are sustainment supply classes currently or recently handled by the DoD distribution network and with the supply chain largely managed by DoD personnel. A few examples in this report also include classes I (subsistence) and VIII (medical materiel).

chain affecting a large number of processes over several years, creating inefficiencies and performance problems as each change was made in isolation.

Review of DoD Supply Chain Policy

The study reviewed the 2003–2004 DoD supply chain policy and regulations in effect when this study was conducted and when the case studies occurred.[3] The review suggested that gaps in supply chain integration have been rooted in DoD supply chain policy. During the writing of this report, from December 2011 through May 2012, though, a new DoD supply chain materiel management policy instruction, informed by this study, was released and the accompanying detailed manual was in the coordination and release process. Both of these were reviewed in the course of the study as well. However, to help illuminate some of the underlying factors and thinking that has hindered supply chain integration and produced the opportunities for improvement discussed in this report, we list the major gaps that have existed in policy:

- absence of an overall supply chain objective that integrates readiness and total cost
- lack of an overarching supply chain framework that clearly articulates the roles of each organization and how each process or function affects the others
- overemphasis on customer responsiveness and inventory minimization versus total cost and meeting customer needs by employing the best standard approaches
- limited linkage of stock positioning to minimize total supply chain costs by integrating inventory, materiel handling, and transportation planning
- limited guidance on when to use different distribution methods, which integrate transportation, materiel handling, and stock positioning planning
- no requirement for collaborative planning with suppliers to enable better management of lead times, order quantities, and costs.

The authors provided overarching and detailed recommendations to address these gaps and to add new policies to engender supply chain integration, in addition to making specific recommendations for the new policy documents. All but the second have already largely been addressed in the new policy instruction and the draft policy manual.[4]

A DoD Supply Chain Framework

The supply chain objective and principles lead to a framework for the DoD supply chain that can provide a common understanding of the design, the roles of each function and process, and dependencies to factor into planning and decisions. Each function and process in the framework has defined, dependent roles as shown in Figure S.1. The framework is described in depth in Chapter Four.

---

[3] Deputy Under Secretary of Defense for Logistics and Materiel Readiness, "DoD Supply Chain Materiel Management Regulation," DoD 4140.1-R, May 23, 2003; Deputy Secretary of Defense, "Supply Chain Materiel Management Policy," DoD Directive 4140.1, April 22, 2004.

[4] Under Secretary of Defense for Acquisition Technology and Logistics, *DoD Supply Chain Materiel Management Policy*, DoD Instruction 4140.01, December 14, 2011; Assistant Secretary of Defense for Logistics and Materiel Readiness, *DoD Supply Chain Materiel Management Procedures*, DoD Manual 4140.01, Volumes 1 through 11, draft as of March 2012.

Figure S.1
DoD Supply Chain Functions and Processes

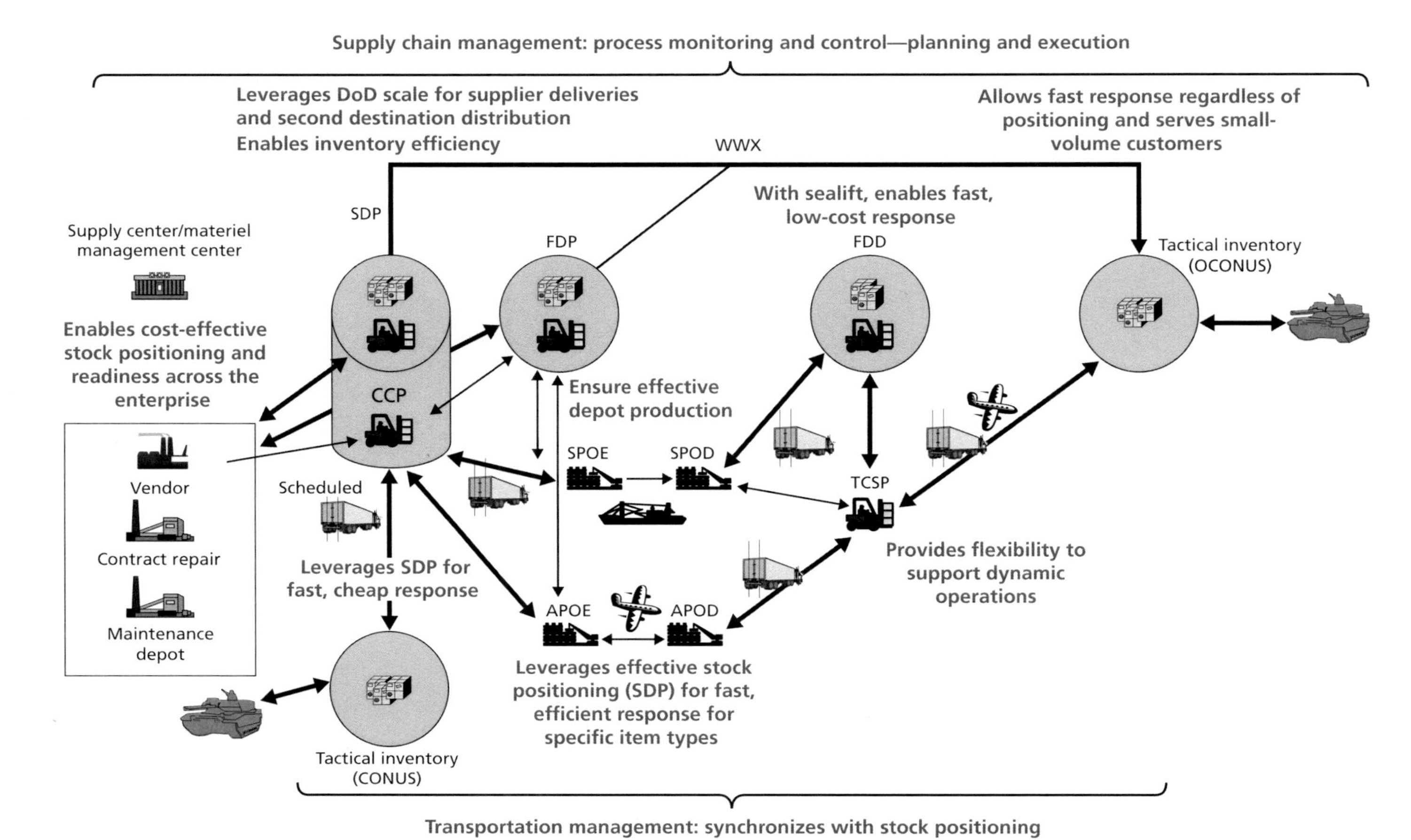

NOTE: All abbreviations can be found in the Abbreviations List.
RAND *TR1274-S.1*

- Tactical and/or retail inventory enables readiness to conduct operations and to execute depot production as planned. Responsive replenishment to tactical and/or retail inventory locations is provided through several means guided by condition-based rules depending upon the customer type, location, and item.
- Overseas, with sealift, forward distribution depots (FDDs) provide low-cost, responsive support for certain types of items. Airlift provides the lowest cost, responsive option for other items. The best type of airlift service depends upon the region and its level of security and development, location demand-level, and item.
- Strategic distribution platforms (SDP)—the distribution hubs in the continental United States (CONUS)—enable low-cost transportation to customers. They also enable lower-cost order fulfillment for suppliers by aggregating regional demand.
- Scheduled trucks in an overseas theater or CONUS provide responsive inexpensive transportation from an FDD or SDP, respectively, leveraging the value of concentrated stock positioning.
- Forward distribution points (FDPs) collocated with industrial activities ensure these activities have the parts on hand to execute planned production.
- Supply management organizations ensure stock is at the right places to take advantage of distribution system economies of scale, when appropriate, or concentrate inventory of expensive items when that is the best solution to minimize total supply chain costs, and they ensure the system has enough—and just enough—inventory to meet service level goals and execute the stock positioning plan. Additionally, they work with suppliers to minimize total costs to meet needs, considering item prices, lead times, order quantities, and quality.
- Transportation management keeps the transportation plan synchronized with stock positioning and ensures responsive delivery upon demand, using the lowest cost options that meet customer needs.
- Overall supply chain management keeps all of these capabilities tied together in both planning and execution. In planning, it ensures all of the dependencies are considered to produce the best overall supply chain design, monitoring the system to determine when plans should shift. In execution, it conducts process monitoring and control to ensure processes are being executed to standard and plan.

Enabling Mechanisms

Enabling mechanisms are management and other approaches that engender execution in accordance with policy and planning intent. They should be reviewed to ensure effective policy execution and to gain the maximum benefit from supply chain integration initiatives. They include the following:

1. Incentives to act in a way that is best for the total supply chain, including metrics to understand individual process and functional effects on the total supply chain and other processes and functions as well as budget accounts and lines that enable and encourage people to take the best actions for the total supply chain.
2. Decision rights and authorities that create spans of control or influence that support integrated action.
3. Decision support tools that enable people to understand the total system effects of their decisions.

4. Financial controls and methods that ensure effective resource stewardship without impeding supply chain efficiency.
5. Information systems that ensure the requisite data for the decision support tools are available and shared.
6. Career development that imbues people with the knowledge and capabilities to act in the best interests of the total supply chain both in formal planning and in ad hoc decisionmaking.

Opportunities to Improve DoD Supply Chain Integration

The study identified several opportunities for improved DoD supply chain efficiency through improved integration:

- Improve supplier management and integration of suppliers, supply planning, and procurement to reduce inventory costs.
- Consolidate shipments in accord with the best systems view.
- Integrate supplier and transportation management with the best systems view.
- Base stock positioning and repositioning decisions on total supply chain costs.
- Integrate financial policy with distribution system design and inventory planning and integrate inventory management across organizations.

Improve Supplier Management and Supplier, Supply Planning, and Procurement Integration

To dramatically reduce supply chain costs, it is critical for DoD to attack the cost paid for material and inventory. The cost of material is the largest element of supply chain costs, and contributing to this, DoD has greater inventory on hand than expected based upon inventory theory, inventory planning parameters, and special categories of inventory requirements unique to DoD, such as war reserve materiel. A DoD-wide inventory stratification report for September 2009 suggests an on-hand "should-be" value of $42.1 billion, with $97.8 billion on hand.[5] This greater-than-expected amount on hand comes from several factors, particularly forecast error accumulated over time and the fact that much of the inventory is in reparable items, which are slow to "wash out" of the system.

DoD forecast error is driven by long lead times, not the quality of the forecast methods. Large order quantities compound this effect as they increase the amount of potential excess when demand diverges from the lead-time forecast. Thus, DoD should begin a new initiative to examine how best to reduce lead times and order quantities, along with item prices. This should encompass how DoD selects, manages, and collaborates with its suppliers; demand and supply planning practices; and organizational design, capabilities, and accountabilities.[6] In addition, DoD should examine how the service materiel/system commands could improve

[5] This excludes U.S. Army Materiel Command and U.S. Army Communications-Electronics Command inventories, which were not included in the stratification report because of an information system transition. It includes all DoD inventory, whether held in the Defense Logistics Agency (DLA) distribution centers or other locations. DoD inventory stratification report, September 2009.

[6] The Assistant Secretary of Defense for Logistics and Materiel Readiness (ASD(L&MR)) initiated such a study in April 2012 to be focused on DLA, which manages most DoD consumable items.

how their demand and supply planning organizations work with depot maintenance and DLA, financial planners, and operational planners to reduce the need for reparable item inventory and new buys, which is essential to having a dramatic impact on DoD inventory.[7]

Consolidate Shipments in Accord with the Best Systems View

DoD also has an opportunity to achieve savings through better integration of its distribution network. We recommend new policy based upon the support provider choosing the best method of service that meets operational needs: When route volume supports well-utilized scheduled trucks that meet these needs, all customers on the route will have their shipments on the truck.[8] To implement such a policy, a central planning organization with the right systems view would need to determine the optimal route structure using an automated route planning tool on a periodic basis.[9] Effective execution is necessary to ensure that delivery standards are met and efficiency expectations are achieved, requiring metrics for monitoring and control.

Integrate Supplier and Transportation Management

Currently, for classes II, IV, and IX (for items stocked in DLA distribution centers), there is no coordination across suppliers to consolidate shipments and suppliers cannot take advantage of DoD's transportation contracts, which would likely be valuable for smaller suppliers. Together these two issues present a potential opportunity for improved transportation, procurement, and supplier management integration for potentially lower total supply chain costs. A rough analysis suggests annual savings on the order of $10 million. Additional savings would be possible if some DoD suppliers are paying higher shipping rates. To achieve these savings, DoD would likely need changes in the Federal Acquisition Regulation to allow for DoD management of inbound freight.

Reposition Materiel Based on Consideration of All Supply Chain Costs

DLA employs a hub-and-spoke distribution network, with regional distribution hubs replenishing distribution center spokes that support depot maintenance operations and overseas forces. DLA was not moving materiel among distribution centers using total supply chain cost logic but rather focused on minimizing some functional costs at the expense of others. A DLA team was formed in the fall of 2011 to examine how to address this issue. The team completed its work in December 2011, with recommendations quickly leading to changes in the DLA business logic for repositioning stock in February 2012.

The new logic minimizes total costs by simultaneously considering inventory, materiel handling, transportation, and procurement workload costs. The conceptual total cost logic can be further extended to DLA stock positioning planning and to service stock positioning and redistribution planning, offering further opportunities. In February 2012, DLA initiated an effort to determine how to apply the concepts to stock positioning.

---

[7] The ASD(L&MR) initiated another study in April 2012 to examine depot-level reparable item management, encompassing an examination across all four services.

[8] This policy recommendation has been incorporated into DoD Manual 4140.01 (draft as of March 2012), which calls for DLA to develop scheduled truck networks based on the principles described here and to make their use standard practice, with exceptions only in accordance with policy guidelines.

[9] In 2012, DLA distribution initiated a project with RAND to transfer the scheduled truck network planning code described in Chapter Seven to DLA for use in a production environment.

Integrate Financial Policy with System Design and Inventory Planning

All DoD operating activities supported by retail and/or tactical supply organizations generate serviceable returns. So the question becomes how serviceable returns can be managed most effectively. It makes sense to keep actively demanded local excess in place to be drawn down. Otherwise, it should be sent back to a central point for reuse. For DLA-managed items, the services transfer money to DLA when they receive this materiel. Often, though, credit for a return is not offered. So the service keeps the item in its inventory, enabling it to reissue the item to the next customer without a second expenditure. The consequence of this practice is some redundancy in distribution system capabilities and masking of demand for DLA planners. There are two potential solution paths to this problem. The first is to ensure that information on service retention stocks of DLA-managed items is integrated into DLA planning systems. The second is to change the credit policy so that there is no incentive to the services for keeping retention stock of DLA-managed items. Changes in the management of retention stock that would eliminate shadow distribution and warehouse capacity would build on a broader DoD trend of rationalizing distribution center capabilities and warehouse capacity.

Conclusions and Overall Recommendations

DoD can increase the integration of its supply chain by addressing shortfalls in policy, enabling mechanisms, and workforce knowledge. Policy creates the foundation upon which to build an integrated supply chain design and the structure within which to work, with enabling mechanisms and workforce knowledge holding it together in the way intended. Fundamental to achieving supply chain integration and pursuing actions consistent with total supply chain optimization as opposed to process or functional optimization is always thinking about doing so, whether in management of the supply chain and its personnel, policy development, process design, and everyday decisionmaking. This starts with ensuring workforce members understand how they affect the rest of the supply chain through a clear DoD supply chain framework—such as the one laid out in this report, receive feedback on their effects on other processes and their effects on the total supply chain, and have the tools to make integrated supply chain decisions. This supply chain framework should be incorporated into DoD supply chain materiel management policy.

DoD has several opportunities to increase supply chain integration with the benefits of improved performance and efficiency. To reduce costs, the most important is increased attention to supplier lead times and order quantities, which can be through increased integration with suppliers. In conjunction, the role of procurement personnel in driving inventory must be recognized to a greater degree. The Office of the Secretary of Defense should launch a new initiative to determine how purchasing and supply management practices could be improved to achieve lead time and order quantity reductions. Related to this is ensuring a tight integration among demand, supply, and repair planning for reparable items to ensure the total supply of unserviceable items in the "closed loop" reparable system is kept to the minimum necessary to support readiness. In 2012, ASD(L&MR) launched two studies to take on these issues with one focusing on improving consumable supply chain management in DLA and one focusing on reparable item management across the services.

Another opportunity is an increased focus on stock positioning, to include improved incorporation of stock positioning in policy and the broad adoption of stock positioning metrics. Improved stock positioning is at the heart of a number of important DoD supply chain initiatives such as Strategic Network Optimization, Distribution Process Owner Strategic Opportunities supply alignment, and the Base Realignment and Closure (BRAC) 2005-based transition to DLA ownership and management of retail stock in support of maintenance depots. It also has an important interplay and potential for leverage with a scheduled truck network improvement effort based upon the scheduled truck chapter in this report. Yet despite the frequency with which stock positioning is the crux of improvement initiatives, emphasis remains limited as reflected in metrics and the lack of goals for stock positioning.[10]

Related to all of these is ensuring that organizations have the breadth of budgets that give them the degrees of freedom to pursue the course of action that will optimize the supply chain and are correspondingly responsible for budgets that they drive the consumption of. A review of supply chain organizational budget categories and the effects that each organization has on costs should be conducted to determine where there is misalignment, with changes made accordingly. Aligning budget authority and organizational effects should also be part of the design process when standing up new organizations or changing organizational designs.

Finally, progress toward supply chain integration could accelerate with improved end-to-end information sharing, to include outside of DoD to the supply base. This includes ensuring each organization knows what information it produces—and more importantly, could produce that it is not—that would be valuable to its upstream and downstream partners. It also includes ensuring that organizations develop capabilities to utilize this information to the full potential.

---

[10] As described in different places in this report, the stock positioning recommendations in this report have been incorporated into the 2012 draft DoD supply chain policy manual, and DLA has incorporated total supply chain cost considerations into OCONUS stock positioning planning and stock repositioning logic and is in the process of revising its CONUS stock positioning logic accordingly. During the course of this study, DLA developed OCONUS stock positioning metrics and goals.

Acknowledgments

First, we thank the Honorable Alan Estevez, ASD(L&MR), not only for sponsoring and supporting this project, but also for engaging in frank, constructive dialogue during in-progress review sessions that both helped shape the direction of and offered opportunities to provide input into other ongoing initiatives. Paul Peters, his Deputy for Supply Chain Integration, provided direct oversight and more frequent guidance along with ensuring tie-in to a policy review and other supply chain integration efforts such as a metrics review, both informing the study and enabling it to have impact during the course of the work. In conjunction, Lieutenant General (retired) Claude V. "Chris" Christianson, the Director of the Center for Joint & Strategic Logistics at the National Defense University, cosponsored the project and provided valuable red teaming and, along with his Vice Director, George Topic, helped synthesize ideas and coordinate meetings. In addition, research on Operation Iraqi Freedom logistics lessons learned sponsored by LTG Christianson when he was Deputy Chief of Staff, G-4, U.S. Army, informed this work and led to a number of related studies for both the Army and joint logistics organizations that further helped crystalize the ideas in this report.

At DLA, a number of people engaged in discussions of data to help draw conclusions, provided data to support analyses, and invited one of the authors, Eric Peltz, into ongoing initiatives to include the updating of the agency performance review metrics, the stock transport order business logic improvement team, the follow-on stock-keeping unit business logic improvement team, the strategic network optimization initiative, and the land and maritime forecasting improvement effort. We appreciate the access and openness to the identification of opportunities to improve defense supply chain operations. In particular, many of the topics in this report involve functions and processes that fall under the purview of the DLA J-33, Materiel Policy, Process, & Assessment. The J-33 Executive Director and his Deputy, Jeff Curtis and Stephen St. John, ensured the requisite data were made available, engaged in discussions to understand the implications of analysis results, and brought in the right people from their organization to assist with these efforts. Within J-33, Robert Carroll, Chief of the Planning Division, brought us into a number of ongoing efforts that fall within his purview, improving our understanding of cross-functional issues. We also want to highlight the tireless efforts he made to bring a large number of stakeholders together and get the theater inventory concepts discussed extensively in this report fully implemented in DLA. Similarly, Teresa Kyte, Chief of Stock Positioning, provided excellent leadership of the stock transport order (STO) business improvement team, enabling rapid implementation of efficiency-enhancing logic now being extended to stock positioning in general. We also recognize the other members of the STO logic team, from whom we gained new insights into the perspectives of different DLA organizations: Jim Laxton (JWL Consulting), Barry Christensen, Joe Habermehl (Habermehl Consulting), Karin Stinson, Tammy Sabo, Randy Gratkowski (Habermehl Consulting), Martha McIntyre, Stephen Rodman, Kimbra Covert, Maritza Guevara, Joel Edgeman, and Richard Davenport. At DLA Distribution, Rear Admiral Thomas Traaen, Joe Faris, and Scott Rosbaugh described their strategic plan and went into great depth on the issues they faced. Brigadier General Darrell Williams, Director of the Land and Maritime Supply Chain, brought us into efforts to better integrate customers into planning and provided an opportunity to engage with his planners and operations research analysts, including James Wagner, Donald Gillespie, Raymond Lowe, Todd Lewis, Anthony Galluch, Cathy Fisher-Farlin, and Roger Decker. Also at Land and Maritime Supply Chain, Captain Roland Wadge, Lieutenant Colonel Andrew DesJardins, Margaret Mickey, Donna Ramsey, Julie Miller, and Pauline Buck provided insight into supplier management practices and challenges. Redding Hobby opened the doors to J-35, where the Strategic Network Optimization team, led by Lynne Allen, provided extensive access into their ongoing efforts, and Lieutenant Commander Manuel Ganuza and Major Chad Ellsworth worked to integrate ideas as the effort evolved. The DLA First Destination Transportation team independently developed the associated ideas in this report in parallel, providing their research and findings and feedback on our efforts. The team included Brett Wood (Analytic Strategies LLC), David Winyard, William Meadows, and John Wait (Analytic Strategies LLC). Colonel Xavier Villarreal, Sean Ahrens, Clifford Lanphier, Rockne Krill, Glenn Starks, Joseph Seawell, John Kurtz, Joy Carter, and Vance Avera also provided information via interviews, responses to data requests, and briefing discussions.

Within the Office of the Deputy Assistant for Supply Chain Integration, we thank Paul Blackwell, Debra Bennett, Carol Conrad, and William Avery (Logistics Management Institute) for providing data, draft policy, and draft metrics and for engaging in policy debate.

Within the Office of the Deputy Chief of Staff, G-4, U.S. Army, Kathy Miller, Colonel (promotable) Duane Gamble, Dave Irvin, Terry Beynon, Rob Saylor, and Lieutenant Colonel David Banian engaged in helpful discussions of integration barriers and opportunities. Dave also played a key role in driving the implementation of the theater inventory concepts in this report.

At U.S. Transportation Command, Vice Admiral Mark Harnitchek (then Deputy Commander), Michael Hansen, Lieutenant Colonel Ralph Lounsbrough, Joseph Lapp, and Brian Campbell provided their thoughts on supply chain integration issues and opportunities and an update on the progress of the Distribution Strategic Opportunities.

We thank Scott Reynolds from the Office of the Assistant Secretary of the Air Force for Installations, Environment, and Logistics; and Maureen Quinlan, Colonel Scott Tew, Thomas Girz, Robert McCormick of the Air Force Materiel Command for providing information on Air Force depot maintenance practices, current initiatives, and the Air Force Global Logistics Support Center and the lessons it offers for supply chain integration.

Additionally, we thank Lieutenant General Kathy Gainey, then Director of Logistics on the Joint Staff, and James Hawkins, her Deputy for Strategic Logistics for offering their thoughts on supply chain integration issues; Lou Kratz at Lockheed Martin for sharing his combined public- and private-sector perspectives; Captain Doug Noble and Captain Robert Brunson of the Supply, Ordnance and Logistic Operation Division in the Office of the Chief of Naval Operations; Jamey Halke and Gus Buhrman of the Naval Supply Systems Command; Brigadier General Kristin French, then Military Assistant to the ASD(L&MR) for coordinating early discussions; and Phil Tombaugh at PRTM for suggesting gaps in what are termed "enabling mechanisms" in this report.

Finally, we thank Lieutenant General (retired) Mitchell Stevenson and Paul Needham, Professor of Logistics and Director of the Supply Chain Concentration at the Industrial College of the Armed Forces for their constructive, helpful reviews of a draft. Over the years, General Stevenson also challenged us to craft DoD-level policy recommendations from our body of research and analysis at RAND.

At RAND, Ken Girardini helped refine recommendations and improve the draft report. Additionally, almost continual interaction in project work with him over the last dozen years contributed immensely to our understanding of the DoD supply chain and how to put theory into practice. Daniel Sommerhauser conducted analyses of scheduled truck improvement opportunities utilizing the scheduled truck optimization program he built for another RAND project, Kristin Klinghoffer mapped inbound freight shipments and produced estimates of cost savings from inbound consolidation. Pat Boren conducted a number of data queries using the Strategic Distribution Database. Daniela Golinelli conducted the statistical analysis of management factors on inventory performance. Ray Pyles offered perspectives on Air Force issues and helped interpret data results for Air Force depot and supply operations. Eric Warner conducted a literature review of reports by and produced for the Department of Defense. Rick Eden helped sharpen briefings and this report. Pamela Thompson and Angelina Becerra assisted with formatting and preparation of the manuscript.

Abbreviations

ALT	administrative lead time
AMC	Army Materiel Command
APOD	aerial port of debarkation
APOE	aerial port of embarkation
ASD(L&MR)	Assistant Secretary of Defense for Logistics and Materiel Readiness
BRAC	Base Realignment and Closure
CAO	contract administration office
CCP	consolidation and containerization point
CIMIP	Comprehensive Inventory Management Improvement Plan
CLIN	contract line item number
CONUS	continental United States
CY	calendar year
DC	distribution center
DCMA	Defense Contract Management Agency
DCRL	Defense Contractor Review List
DDJC	Defense Distribution Depot San Joaquin, CA
DDKS	Defense Distribution Depot Kuwait, Southwest Asia
DDSP	Defense Distribution Depot Susquehanna, PA
DFARS	Defense Federal Acquisition Regulation Supplement
DGPA	Guiding Principles for Acquisition
DLA	Defense Logistics Agency
DLAD	Defense Logistics Acquisition Directive
DoD	Department of Defense

DPAP	Defense Procurement and Acquisition Policy
DS	direct support
DSCP	Defense Supply Center Philadelphia
DSO	Distribution Process Owner Strategic Opportunities
DTR	Defense Transportation Regulation
DTS	Defense Transportation System
DVD	direct vendor delivery
EOQ	economic order quantity
FAR	Federal Acquisition Regulation
FDD	forward distribution depot
FDP	forward distribution point
FDT	first destination transportation
FOB	freight-on-board
FTL	full truck load
FY	fiscal year
GAO	United States Government Accountability Office
GBM	Generalized Boosted Model
GS	General Support
I/A	inspection/acceptance
IPG	issue priority group
IPO	Inventory Policy Optimization
LTL	less than truckload
NDAA	National Defense Authorization Act
NDRI	National Defense Research Institute
NIIN	National Item Identification Number
OA	Obligation Authority
OCONUS	outside the continental United States
OFPP	Office of Federal Procurement Policy
OIF	Operation Iraqi Freedom
OQ	order quantity

OSD	Office of the Secretary of Defense
PLT	production lead time
POL	petroleum, oil, and lubricants
QA	quality assurance
QBO	quantity by owner
RDD	required delivery date
RMC	replenishment method code
RWT	requisition wait time
SAO	Strategic Air Optimization
SASSY	Supported Activities Supply System
SDMI	Strategic Distribution Management Initiative
SDP	strategic distribution platform
SMU	SASSY Management Unit
SPOD	seaport of debarkation
SPOE	seaport of embarkation
SPR	special program requirements
SSA	supply support activity
SS&D	Supply, Storage, and Distribution
STO	stock transport order
SWA	Southwest Asia
TCSP	theater consolidation and shipping point
TDD	time definite delivery
U.S.C.	United States Code
USTRANSCOM	U.S. Transportation Command
WCF	working capital fund
WWX	Worldwide Express

CHAPTER ONE

Introduction

In the mid-1990s, spurred by major shortfalls in logistics processes in Desert Shield and Desert Storm and in the private-sector lean revolution, the Department of Defense (DoD) began a sustained supply chain operations process improvement journey with a substantial emphasis on lean thinking and Six Sigma–oriented programs through initiatives such as the Air Force's Lean Logistics, the Army's Velocity Management, the Defense Logistics Agency's (DLA) and U.S. Transportation Command's (USTRANSCOM) Strategic Distribution Management Initiative (SDMI), and Lean Six Sigma–oriented initiatives in maintenance depot operations.[1] DoD's tackling of new issues that emerged at Operation Iraqi Freedom's (OIF) start and then demands to reduce wartime support costs further fueled these efforts. Rigorous process management, particularly the use of metrics for monitoring and control, became much more prevalent and ingrained in the culture and led to new initiatives, such as the Distribution Process Owner Strategic Opportunities.[2] Much of this was made possible by improved databases and metrics development from earlier efforts—particularly SDMI, along with increasing supply chain visibility with the growing use and effectiveness of radio frequency identification data on shipments. With this sustained business-oriented perspective, the DoD supply chain community has increased performance and harvested significant efficiencies, most notably in the realms of stock positioning to better utilize the distribution network, transportation management, and depot maintenance.

Still, recent analyses and reports indicate that some initiatives offer room for further benefits and that untapped opportunities remain. For example, inventory of repair parts and other secondary items for sustainment is often considered excessive. While there has been significant rationalization of activities within processes and functions across organizations, such as warehousing, accelerated by the Base Realignment and Closure (BRAC) 2005 changes, there appears to have been less progress in integrating the supply chain across functions, both within DoD and with its external partners. One cannot point to existing metrics of supply chain integration or specific performance or cost measures to show supply chain integration

---

[1] For example, see John Dumond, Marygail K. Brauner, Rick Eden, John R. Folkeson, Kenneth J. Girardini, Donna J. Keyser, Eric Peltz, Ellen M. Pint, Mark Y. D. Wang, *Velocity Management: The Business Paradigm That Has Transformed U.S. Army Logistics*, Santa Monica, Calif.: RAND Corporation, MR-1108-A, 2001; Marc Robbins, Patricia Boren, and Kristin J. Leuschner, *The Strategic Distribution System in Support of Operation Enduring Freedom*, Santa Monica, Calif.: RAND Corporation, DB-428-USTC/DLA, 2004; Department of the Army, "Depot Maintenance Initiatives," *2011 Army Posture Statement*, July 11, 2011; Paul G. Kaminski, "Lean Logistics: Better, Faster, Cheaper," speech, Leesburg, Va., October 24, 1996; Richard W. Branson, "High Velocity Maintenance Air Force Organic PDM: Assessing Backshop Priorities and Support," *Air Force Journal of Logistics*, Volume XXXIV, Numbers 3 and 4, June 2011, pp. 16–25.

[2] U.S. Transportation Command, *Distribution Process Owner Strategic Opportunities (DSO) Submission for: Supply Chain Operational Excellence*, 2009.

shortfalls. Rather, this report will show indications of this through patterns of behavior seen in case studies and other brief illustrations, a review of DoD supply chain policy, a discussion of management incentives and examples of remaining opportunities for efficiency that revolve around improving internal DoD integration as well as that between DoD and its suppliers.

This is consistent with findings in the academic literature that achieving supply chain integration—even internally within a firm across functions and business units, let alone externally—is quite challenging, with still limited progress in end-to-end supply chain integration across firms. It involves significant change management hurdles involving new technical capabilities, personnel capabilities and knowledge, organizational goals and incentives, organizational structures, and the nature of relationships, which has a significant impact on the vital information sharing aspect of supply chain integration. Nevertheless, when achieved, both internal and external supply chain integration has been found to have the expected positive effects on logistics and overall firm performance.[3]

To help DoD address the issue of how to become more integrated and tap the full potential of integrated supply chain management, the Assistant Secretary of Defense for Logistics and Materiel Readiness (ASD(L&MR)) asked the RAND National Defense Research Institute (NDRI) to develop a framework for an integrated DoD supply chain, identify barriers and potential enablers to integration, and make recommendations to DoD policy to align it with the framework and engender improved supply chain integration with the intent of reducing costs while ensuring operational needs can be met. In addition, the project also sought to identify opportunities for efficiency through improved integration, building on the findings and recommendations of the first phase of the project.[4] Much of what follows in this report is from an informed perspective based upon a large body of long-term research and direct analytic support involving more than a hundred projects we conducted or oversaw for and with the Army, DLA, and USTRANSCOM; review of ongoing logistics and supply chain initiatives across DoD as of 2011; immersion in several of these initiatives over the last decade; review of recent relevant reports and analyses on the DoD supply chain; interviews and office calls with senior leaders in different DoD organizations; and familiarity with the body of RAND's supply chain management research for the Air Force.

Defining Supply Chain Management and Integration

A *supply chain* consists of all activities involved in getting materiel to a final customer for use, starting with the identification of the need or desire for the item. Supply chain *management* encompasses planning, integrating, and executing these activities to best meet the goals of a

---

[3] The literature on this is summarized in R. Glenn Richey, Jr., Anthony S. Roath, Judith M. Whipple, and Stanley E. Fawcett, "Exploring a Governance Theory of Supply Chain Management: Barriers and Facilitators to Integration," *Journal of Business Logistics*, Vol. 31, No. 1, 2010, pp. 237–256.

[4] The scope of the project was limited to those sustainment supply classes currently or recently handled by the DoD distribution network and with the supply chain largely managed by DoD personnel. These include classes II (clothing, individual equipment, tools, and administrative supplies), IIIP (packaged petroleum, oil, and lubricants [POL]), IV (construction materiel), VI (personal demand items), and IX (repair parts). While there is some discussion in this report of classes I, IIIB, and VIII to provide examples of specific situations, secondary items not included in the scope are class I (subsistence); IIIB (bulk fuel), V (ammunition), VIII (medical materiel), and X (materiel for nonmilitary programs). Class VII (major end items) is not included either.

supply chain's stakeholders. Typically, these activities are executed by a number of different entities that span multiple organizations. For supply chain management to be as effective as possible, then, these organizations have to work in concert to ensure processes and functions are not only synchronized but also integrated. *Integrated* means that they are intertwined in a way that process design and decisions consider impacts on all other processes and the total supply chain in order to achieve optimal supply chain performance and efficiency rather than focusing on the success of individual processes, functions, and organizations.

Achieving Supply Chain Integration

This analysis posits that for the organizations to work in concert as effectively as possible there are a number of prerequisites. The first is a common understanding of the supply chain goal or goals and strategy. In the private sector, it is understood that the goal of each firm is to maximize profit, which leads to the need to find the ideal balance between supply chain costs and service with respect to how the latter affects revenue. This desired balance, in conjunction with the nature of the supply chain in terms of the products and services being provided, should drive the choice of supply chain strategy.[5] For the military, the goal is different from maximizing profit, so the goal first needs to be delineated and commonly understood to determine the strategy. The second prerequisite is agreement on what overall design or structure will best meet these goals and implement the appropriate strategy, given the supply chain's characteristics. The third is clarity on the roles of the different organizations involved in building, operating, and managing this structure and on how they affect each other. As noted in another DoD project, the Joint Supply Chain Architecture, *joint* is often thought of as only when services have to work together. Instead, an interim report from the project notes that to achieve true supply chain integration focused on a common outcome, organizations need to realize that everything they do interacts with the rest of the supply chain and is thus *joint*, requiring each organization to understand how it fits with the rest.[6] The fourth is mechanisms for ensuring that the supply chain works as intended. Given the lack of a supply chain process owner, sometimes suggested as a key to integrating the DoD supply chain, a common understanding of how the entire supply chain should work and the corresponding roles of each organization with incentives to support what is best for the entire supply chain is critical. But one could also argue that even were a single supply chain process owner to be designated, this understanding would be equally crucial given the resulting broad span of control. There would still be just as many functions and subordinate organizations that would have to work in concert. For example, even within DoD's military departments, combatant commands, and agencies and their subordinate commands, there are sometimes competing interests that pursue conflicting objectives due to a lack of overall alignment.

There are gaps with respect to each of these factors that impede the DoD supply chain for sustainment materiel from achieving its full potential. The basic design of the DoD supply chain is sound, but the underlying logic is not broadly understood, leading to inconsistent decisions and partial application of intended practices that create shortfalls in efficiency and

---

[5] Marshall Fisher, "What Is the Right Supply Chain for Your Product?" *Harvard Business Review*, March–April 1997.

[6] PRTM, *DOD Joint Supply Chain Architecture Annotated Briefing of Results and Repeatable Approach Release 2.0*, October 15, 2008.

effectiveness. Additionally, management incentives and practices are not fully aligned with the underlying logic. As per the project's intent, the remainder of this report describes and illustrates the gaps, provides general recommendations for closing them, and offers initiatives to improve DoD supply chain performance that follow from closing these gaps and these recommendations.

CHAPTER TWO

Case Studies That Illustrate the Need for Supply Chain Integration and Systems Thinking

Two related case studies illustrate shortfalls and the need for improvement in DoD supply chain integration. The first case study starts with an example of a supply chain design without an integrated view, but it does culminate in an integrated supply chain solution aligned with the framework for the design of an integrated DoD supply chain laid out later in this report. However, in describing the journey to get to this positive outcome, the case study illuminates gaps in policy, enabling management mechanisms, and the knowledge of the DoD supply chain workforce with regard to supply chain process dependencies and distribution network design principles. The policy gaps it highlights include the absence of an integrating structural framework for the DoD supply chain, the treatment of stock positioning, and the integration of supply planning and stock positioning. With respect to enabling mechanisms, the case study focuses on how functional and organizational barriers created by metrics and budget lines can impede a shift to a more integrated design. Finally, the knowledge gap made change management difficult and could impede future efforts, including optimal execution of the change in design described in this case study.

The second case study illustrates gaps in process integration reflected in information system design shortfalls. It shows the implications when functions and processes are not tightly integrated to ensure changes are coordinated and relevant information is shared and acted on in real time or near real time across the supply chain. In this case, the result was a severe "bullwhip" effect that led to cycles of inventory stock-outs and too much inventory, which were magnified upstream in the supply chain. The case study also demonstrates how people tend to stay within process, functional, and organizational walls rather than bust through them to consider how they could improve the system. In doing so, in conjunction with the first case study, the second case study suggests a need for improved systems thinking in the DoD supply chain workforce so that systems thinking imbues all aspects of supply chain design, interaction, and management. Additionally, the second case study demonstrates how long lead times contribute to the buildup of excessive wholesale inventory when demand on the wholesale supply system declines, leading to long periods of zero orders placed with a supplier. In turn, this makes business more difficult for suppliers as they face "boom" and "bust" cycles at the end of the bullwhip, likely raising costs for DoD.

Theater Inventory Case Study

In fiscal year (FY) 2003, OIF was being conducted with supplemental funding approved in November that did not account for the full pace and scale of operations. In December 2003, with spending outpacing this funding by a wide margin, the U.S. Army began to get concerned that it would exhaust its budget significantly before the end of the FY, forcing very hard choices if some action were not taken earlier. Thus, the Vice Chief of Staff of the Army initiated an effort to identify discretionary spending that could be cut. A resulting memo identified high airlift costs for sustainment as one potential opportunity. This led to proposals to shift airlift to sealift, which would have lengthened delivery times of supplies to the theater, potentially impacting equipment readiness and operational effectiveness. In late 2003 as part of a larger project on OIF sustainment issues and lessons learned, RAND researchers had similarly identified significant sustainment airlift costs as a problem, finding that the vast majority of the airlift cost was concentrated on a small number of relatively inexpensive, big and heavy items (as measured in unit price per unit weight). The researchers further found that this occurred because the centralized theater inventory had few of these big, heavy items that were driving sustainment airlift requirements.[1] At the time, the primary centralized theater inventory of classes II, IIIP, IV, and IX was held in the Army general support (GS) supply support activities (SSAs) in Kuwait in support of Army forces, which constituted almost all of the forces in Iraq in late 2003. Most of the inventory in these GS SSAs came from Army Prepositioned Stocks. With Army sourcing logic, if a unit's direct support (DS) SSA did not have a needed item, the supply system would next check the GS SSAs for this item, before sourcing it from the continental United States (CONUS).[2]

To address this issue, RAND researchers recommended that a greater breadth and quantity of these types of items be positioned in the GS SSAs and replenished via surface to reduce the airlift requirements.[3] Since the items were relatively inexpensive, more inventory could be purchased to cover the longer surface replenishment pipeline and GS SSA safety stock levels for much less than was being spent on airlift. In contrast to some initial inclination to push forward all heavy items into theater inventory, RAND also suggested a more nuanced approach to determining theater inventory, which was the exclusion of relatively expensive items in terms of their ratio of cost to weight. For these items, theater inventory would have dramatically increased inventory costs, outweighing the benefit of reduced airlift costs. To implement this nuanced approach, RAND established business rules to make theater inventory decisions based upon the tradeoff between the avoidance of airlift costs enabled by centralized theater inventory and the increased inventory and double handling costs associated with stocking materiel in the GS SSAs. The Army Materiel Command (AMC) adopted this nuanced approach, quickly adding the top 800 key air transportation cost drivers that met the criteria for theater inventory to the GS SSAs, with some limits based upon national supply

---

[1] Eric Peltz, Marc Robbins, Kenneth J. Girardini, Rick Eden, John Halliday, and Jeffrey Angers, *Sustainment of Army Forces in Operation Iraqi Freedom: Major Findings and Recommendations*, Santa Monica, Calif.: RAND Corporation, MG-342-A, 2005.

[2] In addition, DLA prepositioned a significant amount of class IV materiel (plywood, sand bags, barbed wire, fence posts, and lumber) in Bahrain that was used to support initial operations for all services early in OIF through ad hoc arrangements.

[3] There were three GS SSAs at the time for different categories of items.

shortages stemming from the combination of increased demand and delayed demand forecast increases early in OIF.[4] When DLA stood up Defense Distribution Depot Kuwait, Southwest Asia (DDKS), shortly thereafter in mid-2004, the Army transferred these inventory levels to DDKS, and DLA established initial stockage requirements using analysis by RAND based upon the same concepts. To support a more complete implementation to achieve most of the cost savings potential, this time the analysis identified the top 7,000 additional items to stock in order to minimize total costs.[5]

In July 2004, an additional emergency supplemental was passed, and with increasing acceptance that OIF would not end quickly and would remain a large-scale operation, needed funding became more consistently provided. The urgent need to continue to address sustainment performance shortfalls and ever changing and newly emerging requirements turned attention away from theater stockage. Also, the initial expedient solutions provided sufficient benefit to reduce management focus on improving theater inventory to reduce airlift costs.

Yet the initial expedient solutions based upon the top airlift drivers and low cost per pound items were not optimal. Several issues were present. First, neither the Army nor DLA had a systematic process for updating the theater inventory requirements for the items they manage, with a methodology incorporated into standard planning systems.[6] Second, for DLA-managed items, safety levels to achieve optimal service levels from DDKS were not set. Additionally, for these items, inventory requirements for DDKS did not affect overall global inventory requirements, contributing to replenishment shortfalls and excessive out-of-stock occurrences. And even if they had been, funding was not available to increase total DoD inventory. Third, the stockage criteria and methods for setting depths were not refined to provide optimal solutions. Fourth, DLA stockage was expanded substantially beyond the initial 7,000 items, using the criteria of at least four demands per year to add an item and two demands to retain it, without consideration for balancing the additional costs of theater stockage with the benefit of reduced airlift.[7] Fifth, stockage of some categories of items was not allowed at DDKS. In some cases, this was because of a lack of capabilities at DDKS, such as cutting capabilities for items such as cables. In other cases, though, such exclusions were more arbitrary, with later analysis showing they could be lifted.

The keys for theater inventory, replenished by sealift, are determining when it minimizes total supply chain costs and how much is needed. Filling a requisition from theater inventory enables fast delivery without using expensive overseas airlift, saving money on transportation based upon the cost difference of airlift and sealift, which is used to replenish theater inventory. Achieving this savings on a consistent basis requires a one-time inventory investment to increase the total amount of inventory in the system based upon the optimal safety stock level for the theater inventory and the sealift replenishment time since the materiel in transit in ships is not available for issue. Theater inventory also entails additional materiel handling workload

---

[4] Peltz, 2005.

[5] Eric Peltz, Kenneth J. Girardini, Marc Robbins, and Patricia Boren, *Effectively Sustaining Forces Overseas While Minimizing Supply Chain Costs: Targeted Theater Inventory*, Santa Monica, Calif.: RAND Corporation, DB-524-A/DLA, 2008.

[6] A comprehensive analysis of all DoD items indicated that, when this stockage concept for theater inventory is applied, only a few items managed by the Air Force, the Navy, and the Marine Corps should be in theater inventory (e.g., a total of 27 at DDKS in September 2008). Demand is too low or the item prices are too high for this to be the most cost-effective supply chain solution. Peltz, 2008.

[7] DLA, "Hub & Spoke Operational Business Rules," memorandum, July 21, 2006.

costs since materiel has to be issued and receipted an additional time—from the distribution center (DC) in CONUS to the centralized theater location to the customer instead of directly from the CONUS DC to the customer. When the transportation savings is greater than the attendant inventory and materiel handling costs, theater inventory is the best solution. This applies for lower price-to-weight items with moderate-to-high demand in terms of total weight shipped per year. Higher price-per-pound items are cheaper to fly overseas upon demand, because inventory of these items is expensive. Low-demand, small items are also cheaper to fly overseas because the extra materiel handling costs associated with stocking forward are more than the transportation savings would be. Additionally, items with more consistent demands in theater will have higher inventory turns when stocked in theater than items with highly variable demands, resulting in greater return on investment for the former.[8] The amount of inventory needed—how much additional inventory to buy for items that should be stocked in theater inventory—is determined by the safety level that optimizes the tradeoff among these costs to gain as much transportation savings as possible for as little additional inventory and materiel handling cost as possible. The breakeven point, in terms of the theater inventory service level and thus the safety stock level that balance transportation and inventory costs, can be determined to answer this question.

In expanding stockage, these tradeoffs were not considered; the nuance of selecting some items for stockage and not others based upon trading off these different functional costs was lost. In particular, many very low-demand, light items were added to theater inventory. For these, stockage in theater inventory adds more in materiel handling costs than it saves in transportation. Small items such as gaskets were replenished in quantities as low as one or two and then issued out in the same quantities, increasing materiel handling costs. Shipping them directly to units from CONUS DCs with other items would have had virtually no impact on transportation costs (e.g., put the plastic bag with a washer or gasket on a relatively full pallet), so the net result was increased costs resulting from the extra materiel handling touches. In other cases, small, light but expensive items were also added, increasing inventory costs more than the corresponding transportation savings.[9]

With supply chain responsiveness problems corrected and greatly improved supply chain performance in support of OIF, attention began returning to efficiency in 2006.[10] As part of this shift, RAND, in work for the Army and DLA, developed an optimization algorithm for theater inventory using standard DoD data sources that traded off the additional inventory costs and the additional materiel handling costs against the reduced transportation costs. The algorithm precisely identified what items to stock, the optimal safety levels for each such item, and what items not to stock. Initially, this drew significant interest from the Army and DLA but did not result in action, likely for a number of reasons outlined in the next paragraph.

Through a series of briefings, it became clear that part of the reason for the divergence from the original—and ideal—theater stockage concept and the struggle to implement change came from a common gap in DoD supply chain workforce knowledge with respect to distribu-

---

[8] *Inventory turns* is a metric for measuring inventory efficiency, in conjunction with performance metrics and in the context of a specific business environment, calculated by dividing the annual costs of goods sold by average on-hand inventory.

[9] To illustrate, one example was a pressure indicator with a price of $3,682 and a weight of 0.14 lb. With one demand in a year, the airlift savings were $0.66 but the additional inventory holding cost was $589.

[10] Eric Peltz and Marc Robbins, *Leveraging Complementary Distribution Channels for an Effective, Efficient Global Supply Chain*, Santa Monica, Calif.: RAND Corporation, DB-515-A, 2007.

tion network design involving the need to trade off the three costs discussed.[11] Another aspect of this was a broad belief that stock positioning affects response time as opposed primarily to affecting the transportation cost to achieve a response time. In short, there was a widespread misperception that having materiel stocked in theater would create a readiness advantage due to faster order fulfillment due to the closeness of the stocks. In fact, response times by air from CONUS were about the same, with the two options offering opposing advantages in transportation versus materiel handling and inventory costs. The result of this misperception was greatly expanding the stock list, increasing supply chain costs for tens of thousands of items. Conversely, not being clear on the value of theater inventory in terms of cost, safety levels were not established and incorporated into overall DoD inventory requirements for some of the big, heavy items that drive transportation cost, with replenishments of these items on an opportune basis rather than trying to achieve a cost-optimal service level, which also increased costs. This stems in part from another knowledge gap. In DoD, safety levels are often considered as only being tied to readiness and customer service from a global perspective, rather than also having a role in ensuring cost-effective stock positioning, with local stock-outs sometimes having substantial costs even when there is material elsewhere in the system to meet readiness needs. The extra costs come from excessive transportation costs, particularly for bigger, heavier items. Finally, there was no measurement of total supply chain costs and how the various elements contributed.

While implementation of the recommendations to improve DDKS was stagnating, high airlift costs, which were perceived to be quite excessive, caught the attention of USTRANSCOM, DLA, and AMC, with the commanders jointly calling for reduction in airlift through improved theater inventory and better alignment of shipment priorities with shipping modes in what came to be called the "11-star" memo, as the result of its being signed by the commanding general of AMC, the commander of USTRANSCOM, and the director of DLA.[12] This memo and the increasing attention to efficiency that drove it may have also been spurred in part by an increasing role played by USTRANSCOM in seeking DoD-wide distribution integration and efficiencies reinforced in 2006 by the redesignation of USTRANSCOM as the DoD Distribution Process Owner.[13]

After a series of briefings to senior leaders in the three commands, DLA agreed to implement the optimization concept with respect to the items to add to DDKS, with the exception of classes of items for which exclusions remained, and to set their safety levels and thus depths. DLA established an off-line, monthly process for updating these requirements. Removing items from theater inventory that should not be in theater inventory was not included. AMC also updated its theater inventory but did not establish a standard updating process.

[11] Consistent with this observation through a survey of military and civilian personnel in supply chain management positions, RAND found that the largest shortfall in Army supply chain management knowledge fell in the realm of distribution network design, and this theme has also been observed in joint meetings involving recent initiatives (Thomas Held, Lisa Colabella, Matthew Lewis, John Halliday, and Christopher McLaren, "An Assessment of Opportunities for Improving the Education and Career Development of Army Supply Chain Managers," unpublished RAND Corporation research, 2007).

[12] Lieutenant General Robert T. Dail, General Benjamin S. Griffin, and General Norton A. Schwartz, "Transforming Priority Requisitions to Optimize Distribution," memorandum, October 12, 2006.

[13] Gordon England, "Redesignation of the Commander, United States Transportation Command as the Distribution Process Owner (DPO)," memorandum, May 8, 2006.

However, despite this process change by DLA in 2009, with partial pilot implementations in 2007 and 2008, performance improved only marginally, staying significantly below what RAND calculated as the optimal level and optimal cost savings. Still this was enough to save roughly $200 million per year.[14] The major shortfall was not adjusting the overall system safety level and thus total inventory requirements to account for site-specific service levels, to include DDKS. This led to insufficient stock at times in CONUS to replenish DDKS, leading to significantly lower than optimal service levels. At the time, the DLA Inventory Optimization Model determined a global safety level necessary to meet global materiel availability or service level goals, with the safety level allocated to strategic distribution platforms (SDPs) only. In contrast, meeting overall service level goals and facing fill goals by location would typically require some increase in the total system safety level. So replenishments remained more opportune than fully funded and planned to meet the desired service levels.

Why was this the case? We hypothesize it was because of policy gaps and associated problems in incentives in the forms of metrics, goals, and budget boundaries. Policy has long focused on minimizing global inventory investment to support materiel availability objectives based upon achieving readiness goals. It has not called for ensuring orders are filled from specific wholesale locations to minimize total supply chain cost and correspondingly computing the total system inventory requirements necessary to minimize total supply chain cost. Accordingly, supply organizations have been responsible for materiel availability and inventory costs. They do not have metrics and goals for stock positioning, which they affect and in turn affects total supply chain costs, and they do not have metrics to provide them with feedback on their effects on other costs, particularly transportation. Similarly, their budgets have traditionally been designed to meet materiel availability targets at minimum inventory cost, with DLA and service working capital funds used to purchase inventory. The services have had to pay for outside the continental United States (OCONUS), second destination costs through operation and maintenance budgets, so their operation and maintenance accounts and bill payers receive the budget benefit of theater stockage. Third, for years there has been significant pressure by Congress, the Government Accountability Office (GAO), and DoD financial managers to reduce DoD inventory based upon perceptions of poor management. While these perceptions likely have had some basis for validity and there are opportunities to reduce inventory, these pressures created an environment where any inventory increase was considered bad, even when such an increase would reduce other supply chain costs to a greater degree, lowering total costs. With this combination of factors, supply managers had no incentive to invest in inventory to improve theater inventory and reduce total supply chain costs. Rather, their incentives were the opposite, with such investments risking potential criticism when inventory was looked at in isolation. And even if they wanted to invest in inventory to improve theater stocks, they did not have the budget authority to do so.

As an example, we show the source of fills and material availability for one high-demand item in Southwest Asia (SWA) in Figure 2.1. The lower, yellow series of the sand chart shows orders filled from theater inventory by month, with the middle, dark grey series showing orders filled via airlift from CONUS, and the top, light grey series showing orders filled using direct sealift from CONUS to customers in SWA. This item, with very high volume and a very low price per pound—an ideal theater inventory item—was stocked at DDKS, but for this period

---

[14] Peltz et al., 2008.

Figure 2.1
Source of Fills for SWA Customers for a Sample Item

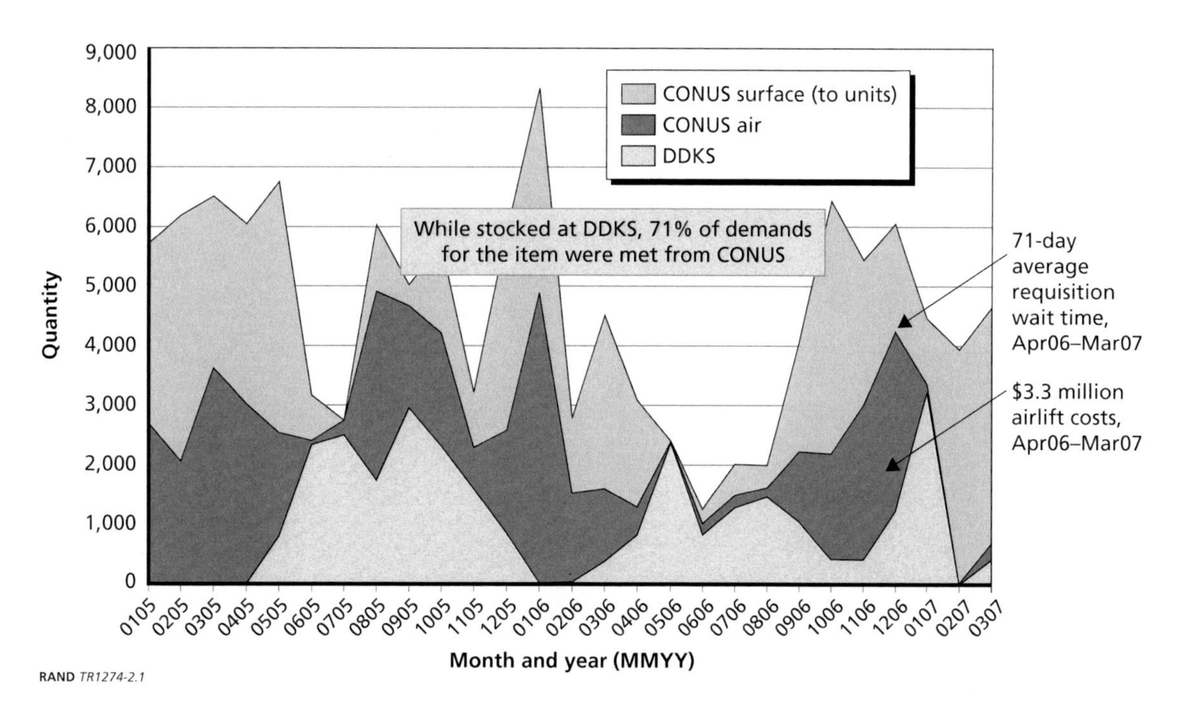

from January 2005 to March 2007, 71 percent of demands were met from CONUS. In some periods, most of the shipments were sent by air at substantial costs. For example, the heavy use of airlift in late 2006 cost $3.3 million. When these shipments were sent by sealift, wait times became excessive, leading to tactical inventory stock-outs, in turn leading to not mission capable vehicles, and likely sometimes leading to maintenance shops double ordering. However, through this period, materiel availability was 100 percent in 36 of 37 months. From a supply or inventory manager's standpoint, this item was performing well—almost always available with limited inventory on the shelf for high inventory turns and low inventory costs. Yet, only a couple hundred thousand dollars in inventory holding costs per year would have been needed to avoid millions in airlift costs and to avoid long wait times and readiness problems from substantial amounts of sealift-based shipments to customers.[15]

Further impetus to address the remaining theater inventory shortfall, driven by not funding increased safety levels and the disincentives opposing any inventory increases, was provided by the adoption of the theater inventory optimization concepts in the USTRANSCOM Distribution Process Owner Distribution Strategic Opportunities under what is called the Supply Alignment initiative in 2008. DLA expressed agreement with the concept but indicated an inability to execute due to a lack of an increased inventory budget. To address this in 2010, the Army transferred about $20 million in Army-owned retention inventory of actively demanded DLA-managed items to DLA at no cost, with DLA agreeing to invest this amount in DDKS safety stock, with the sales of this transferred inventory ultimately replenishing the funds.[16]

[15] Peltz et al., 2008.

[16] Retention inventory is inventory held in SSAs above required DS levels. It could be there as the result of customer returns or reductions in inventory requirements. The Army held this stock to serve customers rather than turning these items in to DLA for very little financial credit and then having to buy them at full price again.

This was only part of the need for DDKS and other OCONUS forward distribution depots (FDDs), though. In 2011, to enable full implementation at DDKS and in all other OCONUS FDDs, USTRANSCOM provided the remaining $40 million using Transportation Working Capital Fund dollars, counting on reduced airlift needs from the improved theater inventory. The attendant safety stock investments were projected to save about $75 million to $100 million per year.[17]

In parallel with these financial decisions, DLA started rolling out the Inventory Policy Optimization (IPO) tool, which went live in January 2010. IPO enables the setting of safety levels by inventory location or stock keeping unit based upon location-specific service goals. This was implemented as part of other system changes in order for DLA to be able to accomplish its new responsibilities resulting from the BRAC 2005 Supply, Storage, and Distribution (SS&D) reconfiguration. DLA's new SS&D responsibilities put DLA in a DS role to industrial activities, requiring specific service levels to conduct operations. Besides ensuring effective support at these activities, the IPO functionality extends to FDDs.

With IPO implemented and sufficient funding secured from the Army and USTRANSCOM, DLA fully implemented the concept in all of its FDDs in July 2011 and added the safety stock requirements to overall inventory requirements, initiating the needed buys in mid-2011, with benefits beginning a procurement lead time beyond these buys.

Notably these concepts came up again in the DLA-led DoD Strategic Network Optimization effort in 2011. Modeling global demand and stock positioning to identify potential opportunities for changes in the network structure and its utilization, the effort identified the major cost savings opportunity as further optimization of stock positioning in OCONUS FDDs to ensure the optimal mix of airlift and sealift. Given that the effort employed 2010 data, prior to the full DLA implementation of full concepts in all FDDs, this result was consistent with the series of analyses identifying the need to integrate inventory, transportation, and materiel handling planning. With the mid-2011 implementation of safety levels, achieving some of the remaining savings potential should be under way with the rest a matter of fine tuning execution with respect to DLA-managed items. Additionally, it should be noted that as of the writing of this report, AMC does not yet have a systematic process for updating its theater inventory of Army-managed items held in the same FDDs.

Supply Chain Silo Case Study

The first case study referred to the 11-star memo, which called for a reduction in airlift through two routes that would better implement existing policies and practices. The first was to improve theater inventory in SWA with the right items, using inventory with sealift replenishment to cost-effectively substitute for airlift for the appropriate items. As described in the previous section, stocking an item overseas is less costly than airlift when the additional inventory cost and materiel handling cost of additional touches (one extra issue and receipt) costs less than the extra cost of airlift over sealift. The second improvement the memo called for was to ensure low-priority requisitions were shipped via sealift rather than airlift as per standard guidance.[18]

---

[17] Based upon RAND and DLA estimates.

[18] Dail, Griffin, and Schwartz, 2006.

Beyond the standard guidance, the services each have specific airlift policies that determine when to use airlift and sealift in support of their units.

The memo resulted in close tracking of airlift drivers, with review by USTRANSCOM, DLA, and the services, using what was called the top-100 list. This listed the top 100 items in airlift cost. The weakness was that it did this regardless of service airlift policies and whether an item should be stocked forward or flown overseas. Despite the call in the memo to cut airlift in cost-effective ways, through a series of misunderstandings by well-intentioned personnel, this actually resulted in the automatic shipping by sealift of a set of items necessary for ground vehicle readiness when these items had to be shipped from CONUS due to theater inventory stock-outs. All of these items should have been stocked overseas but for various reasons were in short supply in DDKS. Since they were readiness items, the services normally called for them to be shipped by air from CONUS when not stocked or out-of-stock OCONUS. Instead, the incorrect use of sealift degraded readiness, potentially increased costs for the total supply chain, and produced a lesson on the problems that can result from supply chain functional and organizational silos, providing a clear example of the bullwhip effect and what happens when actions are not coordinated along the supply chain. We use it to illustrate the resulting problems, with a series of graphs showing the effects of a series of actions with respect to one of these parts.

The first problem, at the heart of the first case study, was that the concept of how to effectively use theater inventory, was not fully understood and implemented by DoD supply organizations at the time. As a result, in some cases, they did not stock the right things or set theater service levels and associated safety stock levels to minimize total supply chain costs. In others, they did not account for theater inventory requirements in global supply planning, sometimes resulting in insufficient supply to adequately replenish theater inventory.

As a result of low wholesale inventory in the system, for the case study part, there were periods of either low or no replenishments to DDKS in 2005 through mid-2006 (see graph 1 in Figure 2.2), leaving it completely out of stock at DDKS most of the time. In this case, the item's reorder point and safety level for DDKS was also set too low, potentially contributing to insufficient total inventory. This stemmed from the lack of a systematic process for updating economically driven DDKS inventory levels. Thus, most shipments to SWA were shipped via airlift, resulting in excessive transportation cost versus the best supply chain design and execution for this part. In graph 2, the blue area shows airlift shipments, the yellow indicates direct sealift to customers with long wait times, the light green depicts DDKS-originating shipments, and the black shows shipments from Army retention stocks in SWA.

To cut the high airlift cost seen in graph 2, in late March 2007, shipments from CONUS were switched primarily to sealift, as shown in graph 3. Accordingly, the wait time for customers in SWA increased (see graph 4). This switch was not communicated to other functions and organizations in the supply chain. Thus, this wait time increase was a surprise to customers in SWA and their tactical inventory and supply chain planners. Consequently, they did not have a chance to decide to increase their tactical inventories in advance to accommodate longer replenishment times, resulting in rapidly increasing backorders to their maintenance customers as shown in graph 5. They had been receiving the part in a little over 10 days, with inventory planned to accommodate 20 days or less, and expected this trend to continue. Given the lack of notification of a change in shipping mode and DoD's lack of specific estimated delivery dates for orders, the new wait times were not clear until the first sealift-based shipments actually arrived.

Figure 2.2
First Seven Events in the Supply Chain Silo Case Study

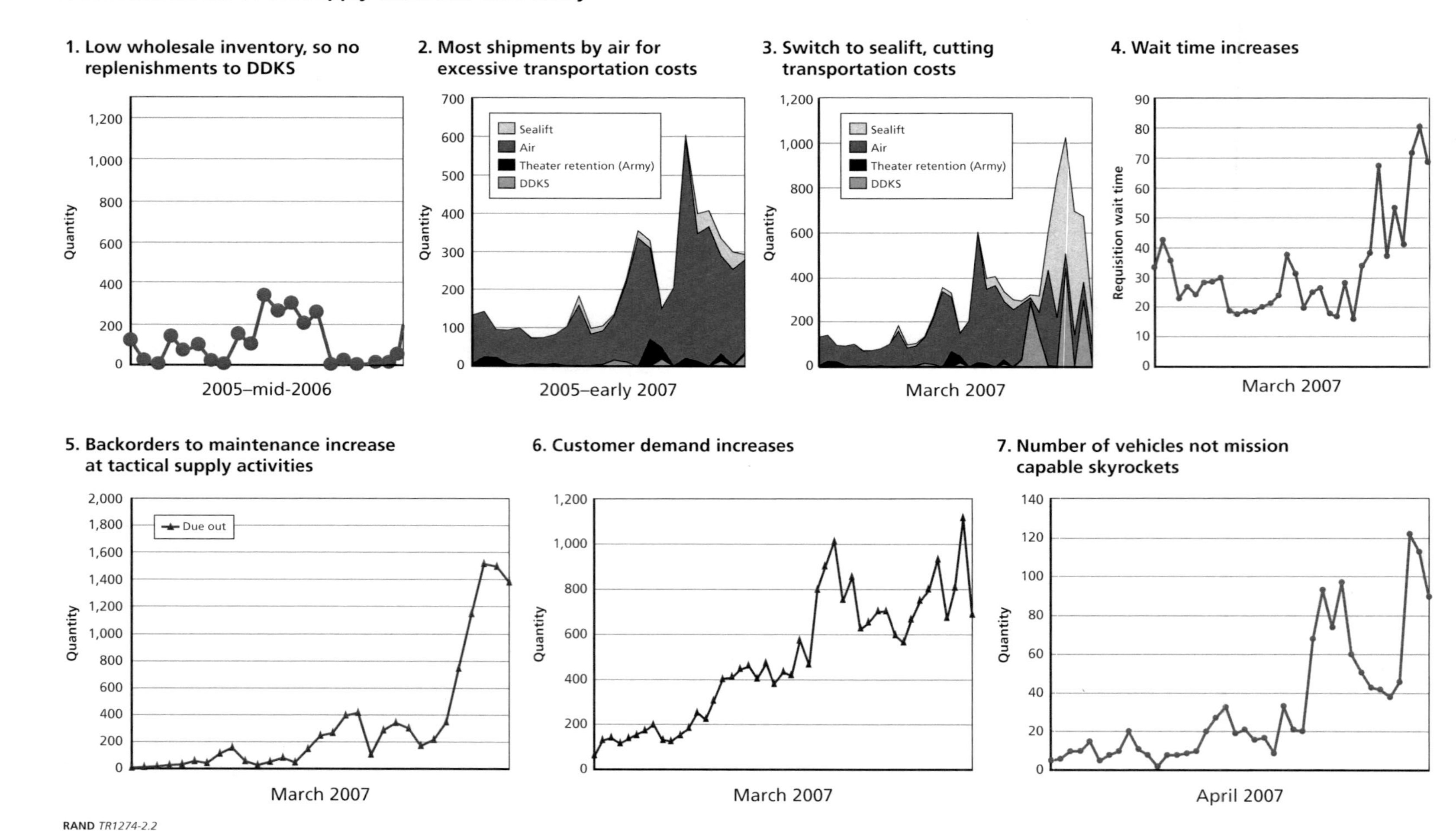

In fact, all that the maintainers would have seen is that their orders were starting to be open for an increasingly long amount of time. At about the same time, there was an increase in maintenance customer demands, as seen in graph 6. This corresponds to both the start of the summer and the 2007 surge in Iraq, so some demand increase is not surprising. However, one sees more than a doubling of demand when sealift shipments start, with this very high level remaining until there is a shift back to airlift. It has been hypothesized by many that when supply chain problems occur, maintainers and other customers sometimes double order, just in case. It is possible that some level of double ordering contributed to the very high demand seen from March through July 2007, after which it dropped down somewhat (it is not possible to determine if a demand is a double order against the same maintenance work order from the available data). In turn, the backorders to maintenance from tactical inventory resulted in substantially increased numbers of vehicles that were not mission capable (graph 7), producing lower readiness, just for this one part. Typically, very few vehicles are down for this part, because all tactical supply activities stock it and provide it quickly upon demand.

When inventory planners executed their next quarterly inventory adjustments, they dramatically increased their inventory requirements based upon the higher demand levels (see the blue line in graph 8 in Figure 2.3). They actually decided not to increase the planned replenishment wait time, which would have further increased the inventory requirements. The primary reason was that many of the tactical supply activities were running short of available storage space. The effect of not increasing the inventory requirement in accordance with the replenishment time, though, was to reduce the planned service level provided by the tactical supply activities. However, this type of non-action is the best course of "action" when an upstream supply chain disruption or problem temporarily increases the replenishment time without warning. Increasing the inventory to match an unplanned, temporary, higher replenishment time would increase the demand signal to the upstream supplier, creating or exacerbating the bullwhip effect. If an increase in replenishment time were planned in advance of execution, ideally, such a planned switch to sealift would have been communicated to inventory planners a sealift-wait-time ahead, with preadjustment of the inventory requirement. This was a service-managed item, so the tactical inventory increases led to a one-time increase in the overall inventory requirement.

It should be noted that had this been a DLA-managed item at the time (this item's management was transferred from AMC to DLA in 2011), the supply activities would have placed orders with DLA to fill these one-time inventory requirement increases. However, the service's orders do not normally include a code to indicate when orders represent a one-time inventory increase, reflecting a non-recurring demand.[19] Additionally, in demand planning, the DLA information system does not currently distinguish non-recurring demands from retail supply activities from other demands from such activities it receives. So the tactical inventory increases would have produced the higher demand signal to DLA seen by the yellow line in graph 9. The black line in graph 9 shows that, for several months, orders to DLA from the supply activities exceeded the incoming maintenance customer demands. Instead of being treated as one-time non-recurring demands, these orders would have been incorporated into demand history files used to produce forecasts of future demand that, in turn, are used to determine both wholesale inventory requirements and procurement plans. This type of issue would have in fact occurred

[19] The information system does have a field to indicate recurring or non-recurring demand, but it defaults to recurring, with non-recurring rarely used, without an automated ability to do so for one-time inventory increases.

Figure 2.3
Events 8 Through 13 in the Supply Chain Silo Case Study

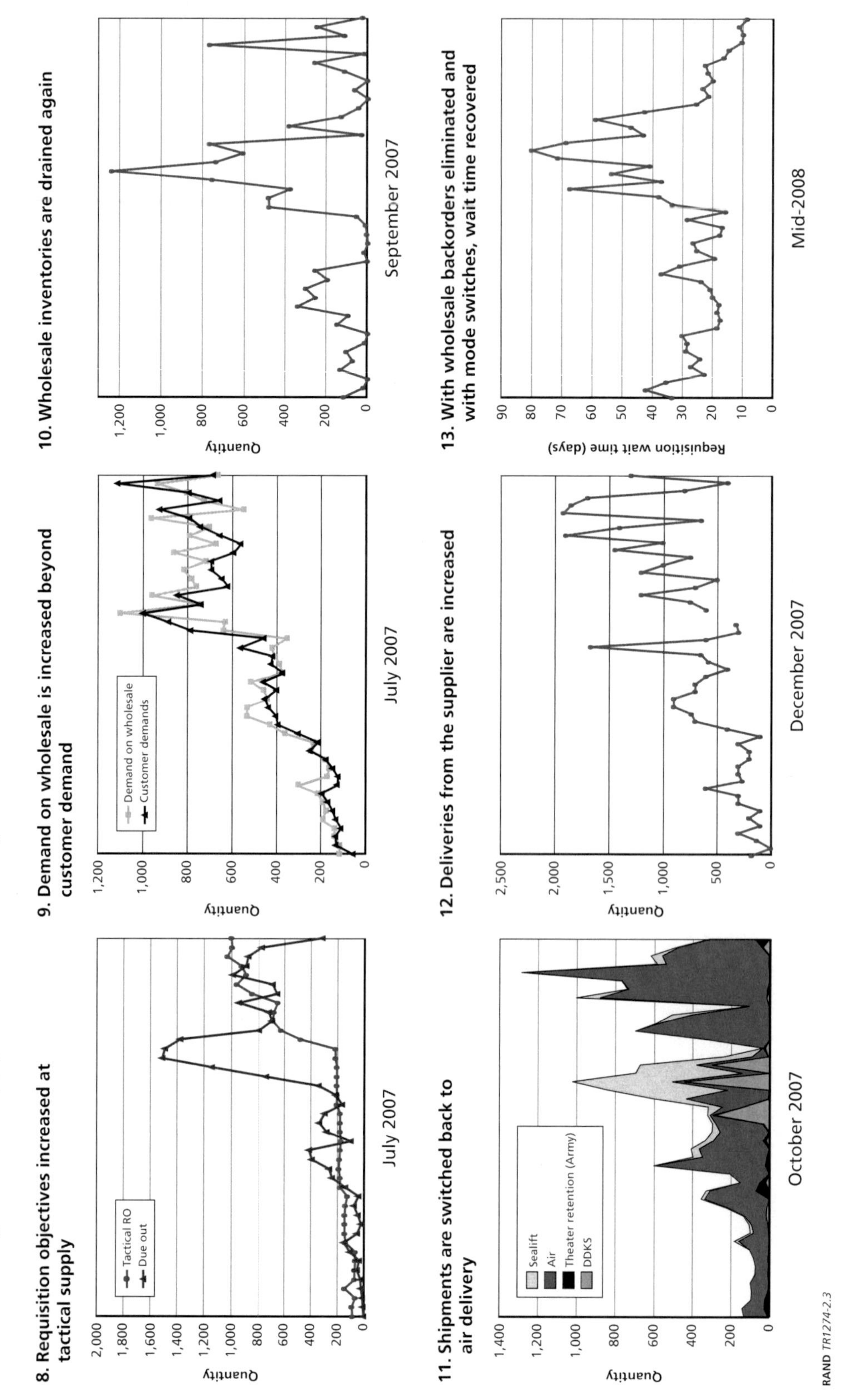

RAND *TR1274-2.3*

for the large number of DLA-managed items stocked in service-managed supply activities, such as the 50 or so new ones constituted in Iraq early in OIF.

Going back to the actual story for this item, since the tactical inventory increase was not pre-coordinated and communicated with the national-level or wholesale planners (and if the demand increase was a surprise, it could not have been coordinated and communicated in advance), total inventory and replenishments were not adjusted in anticipation of the need to increase tactical inventories through issues from CONUS distribution centers; rather, they reacted. So, just when national-level inventory had caught up, these unanticipated tactical inventory increases, in combination with the increase in maintenance customer demand drained the national level-stocks (see graph 10). It is possible that the maintenance customer demand was overstated as well, in reaction to the long wait times and the increase in not-mission-capable vehicles. Thus, the wholesale planner ordered stock to increase total inventory and fill the tactical inventory needs, increased orders to fill a projected increase in ongoing demand, and had to order to fill the increased wholesale distribution center inventory requirement, which would have also gone up in response to the increase in demand. Therefore, the wholesale supply organization dramatically increased orders to its suppliers for multiple months. As shown in graph 12, a procurement lead-time away in late 2007, deliveries from suppliers increased dramatically.

While wholesale was reacting to the signals it was receiving, the readiness problem led to an investigation of the problem and discovery that shipments had been mistakenly switched to sealift. To address the readiness problem, shipments were switched back to airlift in September 2007 (graph 11), with the additional airlift costs accepted as temporarily necessary to conduct operations. Then, with the increased orders placed with the supplier arriving, wholesale inventory recovered, with the combination of airlift and the elimination of wholesale backorders enabling shorter wait times for requisitions to be filled (graph 13).

The recovery at wholesale also enabled DDKS replenishments, so it could begin filling orders (Figure 2.4, graph 14). A sealift lead time after tactical inventory adjustments, the resulting orders arrived to increase inventories. Combined with the shorter distribution times from DDKS and elimination of wholesale backorders, this led to the elimination of tactical backorders (graph 15). As described earlier, if the tactical inventory had been increased to the level needed to accommodate the sealift replenishment time, once the switch back to airlift occurred, tactical supply activities would have had more stock on hand than needed.

As people have hypothesized may happen in response to poor supply chain performance, some evidence has suggested that sealift shipments of readiness drivers do lead to double ordering and workarounds to satisfy maintenance needs.[20] Once wait times and service to maintenance returned back to immediate issues from tactical SSAs, the spike in serviceable returns in late 2008 (graph 16) raises the possibility that this happened in this situation. If so, the double ordering would have magnified the demand signal to wholesale, with the returns then leading to suppression of demands on wholesale.

As demand began to drop, tactical inventories were gradually reduced (graph 17). In contrast to the inventory increase case, if this had been a wholesale item, during this period, the demands on wholesale would have been suppressed while they drew down their inventory. This reduced tactical inventory requirement, which temporarily suppressed tactical inventory

[20] Unpublished RAND research by Ken Girardini.

Figure 2.4
Events 14 to 19 in the Supply Chain Silo Case Study

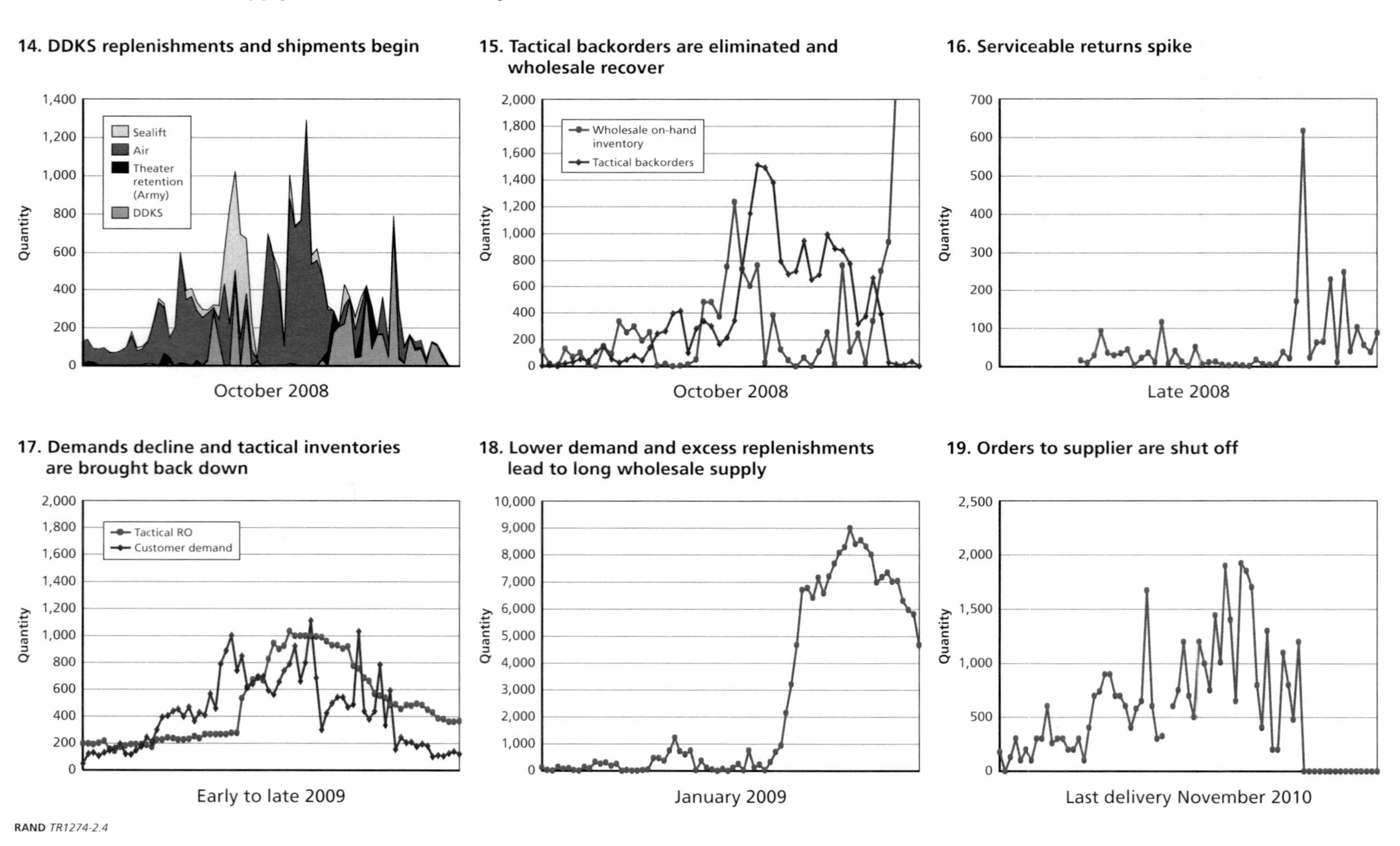

replenishment orders, combined with the decrease in demand to produce significant wholesale inventory excess. Further, orders from maintenance then dropped more as the use of this vehicle began to partially shift to another. Since orders to the supply base have to be placed a procurement lead-time ahead, replenishments to wholesale continued to come in even after this situation was clear to wholesale planners, further increasing the inventory excess seen in graph 18. A lead time after the wholesale inventory and forecast problem became clear, deliveries from the supplier ceased. Orders remained shut off through the end of 2011, with the supplier effectively becoming dormant for this part (graph 19).

Additionally, the supplier may not be ready to respond once orders need to be restarted, producing a longer lead time than expected, or the lead time on the next order could be higher due to factors related to production dormancy or the need to create a new contract. If wholesale assumes the lead time will be as in the past, the startup of orders will be placed too late, resulting in wholesale stock-outs in 2012 or beyond and perhaps restarting the six-year cycle shown here. Finally, we see a pattern of "boom and bust" for the supplier, which likely increases its costs and the prices that DoD pays.

Again, if this had been a DLA item, if the fact that retail orders went down in part due to inventory drawdowns—and not due to a real underlying decrease in demand—was not communicated to wholesale, this would have caused wholesale planning systems and planners to under-forecast demand. In turn, this would have led to projecting that the on-hand inventory would last longer than it actually would once demand on wholesale returned to reflect the underlying, ongoing customer demand level.

The Need for Systems Thinking

We see in both of these case studies that the 11-star memo created impetus for change, but even then this was slow in coming because of incentive and budget misalignments. The memo also inadvertently precipitated the second case study on the need for enhanced supply chain integration through improved communication and data sharing embedded in standard processes. The case studies also suggest the need for a bottom-up approach to supply chain integration through improved systems thinking in which people take supply chain integration into account in the course of the thousands of individual actions that they take every day. The memo said to cut costs in ways that would not degrade readiness—in other words, cut costs smartly. For example, it said: "Utilize cheaper sealift vice expensive airlift when mission requirements allow" and "divert, wherever practical, items to surface . . . . This would also involve an extensive review of items and levels we stock forward."[21]

However, focusing on the need to reduce airlift, some planners automatically diverted a number of critical readiness items to sealift, without ensuring associated improvements in theater inventory. This was done in part due to misunderstanding, but it is also possible that those in control of planning transportation were overly focused on finding ways to cut transportation costs without carefully considering the broader supply chain implications of their actions. Under pressure to produce results through forums such as the top-100 reviews, they would have been responding to the part of the message they could control. Also, they might not have understood the interplay among the different supply chain costs. As discussed in the second

[21] Dail, Griffin, and Schwartz, 2006.

case study, this initial action created readiness problems and likely increased total system costs as the result of a series of uncoordinated actions that led to severe bullwhip perturbations across the supply chain.

As one goes down the chain of command, personnel tend to have narrower, more functionally oriented positions. When personnel are immersed in functional activities all day long, the question becomes "what part of a message will they hear?" Will they fully understand the implications of a directive with regard to cross-functional integration and interactions? What will ensure they execute in accordance with intent and the best overall supply chain or systems solution?

So again, the second case study shows the need to imbue the workforce with a strong understanding of how an integrated DoD supply chain would work and to imbue them with an increased proclivity for systems thinking and taking a systems view when tackling problems and acting. This is only valuable, though, if reinforced by policy, metrics, and budget flexibility. The case study also shows the need to ensure information, plans, and data are shared consistently, through both manual means and automated information system transactions as appropriate, across the supply chain with other functions and organizations, which in turn should have capabilities that enable them to proactively respond to changes elsewhere in the supply chain. This starts with an understanding of what information each process and function needs from all of the others, which starts with an understanding of how they interact and what the dependencies throughout the supply chain are.

CHAPTER THREE

Policy Review

In DoD, policy sets the overall tone by providing goals and guidance that set the bounds within which to operate. A review of policy and regulations that were in effect when this study was conducted and when the case studies occurred suggests that gaps in DoD supply chain integration have been rooted in DoD supply chain policy. As of the writing of this report, *DoD Supply Chain Materiel Management Procedures* (Assistant Secretary of Defense for Logistics and Materiel Readiness, DoD Manual 4140.01, Volumes 1 through 11, draft as of March 2012), which was informed by this study, was in the approval and release process. The rest of this chapter reviews the policy that has been in effect to help illuminate some of the underlying factors and thinking that have hindered supply chain integration and produced the opportunities for improvement discussed in Chapters Six through Ten. In addition, this review, along with a detailed review of the initial draft of the new policy, produced the policy recommendations proposed in the next chapter and to DoD as it refined the forthcoming policy.

First, policy embodied in DoD Directive 4140.1, "Supply Chain Materiel Management Policy," 2004, and DoD 4140.1-R, "DoD Supply Chain Materiel Management Regulation," 2003, lacks a clear articulation of the overarching supply chain objective that integrates readiness and total cost and that succinctly defines what *readiness* means from a sustainment perspective. Second, there are gaps in the guiding principles for DoD supply chain design and decisionmaking. Third, and perhaps most importantly, there is not a delineated, overarching structure or framework that provides a broad understanding of the roles of the major DoD supply chain components or elements, the dependencies among them, and how the individual elements, different functions, and different processes should be integrated. Throughout, policy is written from a process or functional view, without clear articulation of how each process or function affects the others and thus how effects on downstream processes should be considered. The interdependencies among processes are not described, and policies do not ensure these are taken into account. Fourth, in some areas there is no condition-based guidance on when to use the array of different standard approaches or process options, particularly with respect to approaches that require different processes to act in concert and ensure the standardized use of best practices. These could be thought of as rules that, when followed under the specified conditions, would ensure using the best approach for supply chain integration. While policy by itself cannot ensure effective execution, it lays the groundwork for how the system should work, and in terms of supply chain integration, it should describe how the different processes and functions should interact. With policy in place, enabling mechanisms, such as metrics aligned with overall outcomes, can then be used to drive toward optimal execution.

Besides not clearly integrating total costs and readiness in a combined objective statement at a high level, DoD policy has overemphasized customer responsiveness and inventory minimization versus total cost and meeting customer needs by employing the best standard approaches to meet these needs. The DoD "Supply Chain Materiel Management Policy" Directive specifies, among other policies, that

> 3.1. DoD materiel management shall be structured to be responsive to customer requirements during peacetime and war.
>
> 3.2. All costs associated with materiel management, including acquisition, distribution, transportation, storage, maintenance, and disposal shall be considered in materiel management decisions.
>
> 3.3. The materiel management functions shall be implemented with DoD standard data systems.
>
> 3.4. The secondary item inventory shall be sized to minimize the Department's investment while providing the inventory needed to support both peacetime and war requirements.[1]

This requires responsiveness to customer requirements, without indicating limits. And while it directs the consideration of total costs, it calls for minimizing inventory.

These emphases are reinforced in the guiding principles of the supporting regulation, which provides guidance for

> [d]eveloping materiel requirements based on customer expectations while minimizing the DoD investment in inventories. . . . [e]stablishes the customer as the foundation driving all materiel management decision-making . . . [and] encourages the DoD Components to . . . [e]stablish end-to-end processes that are focused on maximizing customer service or warfighter support [DoD 4140.1-R].

The goals echo the directive calling for the components to "provide responsive, consistent, and reliable support to the war fighter during peacetime and war. That support should be dictated by performance agreements with customers to the furthest extent." It then calls for considering total costs but directs minimizing inventories.[2]

While satisfying the customer requirements are clearly important, it is important to also ensure that there are checks and balances on requirements, keeping them aligned with readiness needs and balanced when costs begin forcing tradeoffs. Policy should call for meeting needs and controlling the total cost to do so instead of calling for maximizing support.[3] Additionally, policy opens the door to excessive freedom to customize approaches and performance requirements. Instead, common requirements for the conduct of military operations can be developed, and the best approaches for given requirements and conditions can be identified for standardization around situation-based and condition-based best practices. Overall, there is not a sufficient sense of balance between service and cost or a call for minimizing total costs

---

[1] Deputy Secretary of Defense, "Supply Chain Materiel Management Policy," DoD Directive 4140.1, April 22, 2004.

[2] DoD 4140.1-R.

[3] This is most commonly seen in contractor logistics support and in the aim to meet time definite delivery standards without accompanying these standards with stock positioning standards to ensure the delivery times are met cost-effectively. Similarly, DoD perfect orders do not consider whether the customer orders were met cost-effectively. As illustrated by the second case study, though, this emphasis is sometimes overridden by other factors or incentives.

to meet specified customer requirements, and policy also places minimizing inventory above stewardship of other and overall supply chain resources.[4]

While generally it is valuable to minimize inventory to meet customer service or readiness needs, there are also cases where more than this level of inventory can reduce total costs by enabling less expensive transportation to meet delivery time requirements. Inventory in DoD is typically thought of as being needed to provide readiness through a service-level goal, with the wholesale perspective being on ensuring overall targeted materiel availability without regard for location within an echelon of stock. This is reinforced by treatment in policy, which describes wholesale inventory as one global account, with being able to serve a customer from the overall account the right amount of time to meet readiness needs while minimizing the global quantity being the goal.[5] Policy does not fully address stock positioning within echelons, thus it does not sufficiently prescribe where material should be provided from, which affects the cost to provide the material. In consonance with this, there is not policy on all aspects of stock positioning, fully reflecting its role in total costs and the need to integrate it with overall inventory and transportation planning.[6] The emphasis is on retail versus wholesale, with some discussion on when to use SDPs. When to use OCONUS FDDs is absent; while some places call for incorporating transportation costs in stock positioning decisions, policy calls for minimizing the number of wholesale stockage locations as opposed to determining the optimal number and locations;[7] and incorporating stock positioning in supply planning is not specifically called for. However, inventory levels and positioning can serve as an enabler of different supply chain approaches that reduce total cost for the same level of readiness but with some-

---

[4] The new December 2011 DoD supply chain management policy instruction and accompanying policy manual (in draft form as of the writing of this report) correct this imbalance and emphasize minimizing total costs as opposed to just inventory costs, eliminating these problems. Under Secretary of Defense for Acquisition Technology and Logistics, *DoD Supply Chain Materiel Management Policy*, DoD Instruction 4140.01, December 14, 2011; DoD Manual 4140.01 (draft as of March 2012).

[5] For example, in DoD 4140.1-R, see

> C2.4.2.2.1. For items that are essential to weapon system performance, the inventory performance goals shall relate to the readiness goal of the weapon system throughout its life cycle, e.g., operational availability, mission capable rates.
>
> C2.4.2.2.2. For items that are non-essential to weapon systems or are non-weapon system items, the inventory performance goals shall relate to the time to fill a customer's order, whether that order is a requisition placed on an ICP or a demand request placed on a retail supply activity.

[6] Note that the absence of attention to stock positioning has sometimes led to transportation costs considered so excessive that they have put support to operational activities at risk of being subordinated to cost considerations, with this risk being realized in the second case study.

[7] DoD 4140.1-R:

> C2.6.1.1.4. When possible to achieve weapon system performance objectives, RBS models should be capable of optimizing support across both the wholesale and retail echelons of supply.
>
> C2.6.3.1.6. Safety Level. Due to fluctuations in demand over lead times, repair cycle times, attrition rates, and in other variables, safety level quantities may be stocked as a buffer against backorders.
>
> C5.2.2.1.1. Stocked items shall be positioned to maximize customer responsiveness while minimizing the total stockage, distribution, and transportation costs. Procured items shall be shipped from commercial sources to the DoD geographic distribution depot that provides the best value.
>
> C5.2.2.1.1.2. To the maximum extent possible, stocked items should be positioned so a given customer is supported from the minimum number of wholesale distribution depots and/or other activities.

what more inventory in some cases. Historically, the cost of systemwide backorders has been considered, but the costs of local stock-outs have not been in setting service levels. However, this has started to change with the implementation of integrated theater stockage concepts.

Thus, key recommendations for policy are to enhance the section on stock positioning policy in general to better reflect how to make choices among wholesale locations and how many to use and to modify supply planning policy to reflect the notion that location-specific service levels, based upon minimizing total system cost while meeting readiness needs, should be considered in addition to the overall service level.[8] In other words, both meeting readiness goals through adequate inventory and using inventory to control total costs should be more consistently considered.[9] Further related to these issues is the absence of policy (to include the Defense Transportation Regulation) on when to use different distribution methods, which integrate transportation, materiel handling and shipment preparation, and stock positioning planning. For example, policy does not include when to use scheduled trucks or when and how to consolidate shipments in CONUS for overseas delivery. In short, the policy focuses on individual processes and functions, not how they interact and should be considered jointly for effective integration across the supply chain.[10]

An additional gap in policy is the need for collaborative planning with suppliers to enable better management of lead times and order quantities and even their costs, in recognition of the driving role that supplier performance plays in inventory requirements and overall supply chain costs.[11] Overall, upstream supply chain policy for improved supplier integration could be further enhanced.

Finally, there is a focus on collaborating between providers and customers as a route toward optimization. This creates the potential for excessive system complexity, with highly varying requirements. And it suggests more flexibility and control in performance versus cost than is possible. For example, there may only be three transportation modes that can be used to reach a destination. If each transportation mode is provided with optimal process execution, then the choice of performance becomes one of choosing among the feasible modes for a location and their optimal performance levels rather than being able to choose from among a continuously variable range of performance and presumably cost.

A different perspective on collaboration would be on collaborating between providers and customers for information exchange that enables providers to understand customer needs and then employ the best structure and methods to meet those needs. This draws on the premise that there are some standard best practices and finite sets of options with relatively fixed "should-be" capabilities for each for a given area, theater, or process. In this perspective, the customer would set requirements and then the provider would choose the lowest cost option among this finite set that meets the requirements. In conjunction, policy could call for applying more standard service and/or theater-level readiness and other performance goals, with modifications by exception. This change in perspective would shift the focus of information

---

[8] These stock positioning recommendations have been incorporated into DoD Manual 4140.01 (draft as of March 2012).

[9] Another cost to consider when incorporating total cost considerations into inventory planning should be the impacts of stock-outs on depot maintenance. The question would be whether current depot customer service goals accurately reflect the costs associated with production disruptions and changes.

[10] DoD Manual 4140.01 (draft as of March 2012) largely addresses the policy gap on distribution methods and the need to integrate transportation and stock positioning planning.

[11] DoD Manual 4140.01 (draft as of March 2012) addresses this gap.

exchange to effective support planning. Customers need to continually provide information that affects support planning, and providers need to provide information that is needed for operational planning.

Beyond policy, supporting mechanisms sometimes impede supply chain integration. It is not enough to have policy, whether at the DoD or component level or whether official or de facto as the result of decisions that have been made since the last policy update, to ensure effective execution. In this report, enabling mechanisms are defined as management and other approaches that influence, or even enable, people to act in alignment with policy, which should also drive them to act in the best interest of the total supply chain. The case studies illustrate issues with respect to enabling mechanisms such as the lack of metric-based goals to provide supply organizations with feedback on and accountability for their effect on stock positioning and transportation costs. Similarly, their budgets do not enable tradeoffs between spending on inventory and transportation. Another example would be people not having the span of control or decision rights to manage processes from a total systems view. These examples will be expanded on in a later chapter on enabling mechanisms.

Given the overall findings and these policy assessments and recommendations, this report recommends the development or improvement of the following policy and supporting elements and provides starting points for all but the condition-based guidance, which would involve going into the details of each process (recommended detailed changes to policy on procedures for DoD Manual 4140.01, draft as of March 2012, were provided to DoD as part of the review of the draft policy update):

- a clear supply chain objective integrating readiness and total cost
- supply chain guiding principles
- an overarching supply chain structure that clarifies
  - the roles of the major supply chain elements or components
  - how the major components and the processes should be integrated
- condition-based guidance on when to use the array of standard approaches and options
- enabling mechanisms aligned with supply chain integration.

This list also implies the establishment of a layered framework for developing, reviewing, and refining policy. The layers are

- *guiding principles* for system design and decisionmaking.
- *the structure* to delineate roles and interrelationships of key supply chain components.
- *rules* to provide a condition-based approach for using the structure when a standard course of action applies and ensures chain integration. These rules quickly enable actions consistent with supply chain integration. When the situation goes beyond standard conditions, a business case analysis becomes necessary. These should fall back on the guiding principles, which should engender consideration of total supply chain effects when decisionmaking complexity increases.
- *detailed procedures* for implementing the rules.
- *enabling mechanisms* that enable effective policy execution
  - metrics
  - decision authority
  - financial controls and budget lines

- decision support tools
- information sharing
- workforce knowledge.

CHAPTER FOUR

High-Level Policy Recommendations

This chapter provides recommendations for the development and revisions of DoD supply chain materiel management policy with respect to the overall objective, guiding principles and the integrating structure. To a large degree, they have been incorporated into the draft *DoD Supply Chain Materiel Management Policy*, DoD Instruction 4140.01 (draft as of March 2012), and the supporting manual of procedures that were in coordination at the time this report was written.

The Supply Chain Objective

In the private sector, the objective for the companies in a supply chain is to maximize profit, with the various organizations both competing for shares of this profit and collaborating to maximize total supply chain profit. Maximal profit comes from the combination of service that produces the best combination of revenue (sales and price) and cost to serve. This varies depending upon the type of good or service. Customers are in turn trying to maximize their utility, which is similarly defined as the optimal mix of price and service.

For DoD, the objective is also to find the best combination of service and cost, but service does not translate into revenue through its effect on sales as it does for private-sector supply chains. Thus, the objective function for the supply chain providers in the DoD supply chain cannot be to maximize profit. If one could translate defense capability into monetary value, this one-dimensional objective function would be possible to utilize, but the monetary value of defense capabilities are not commonly agreed upon, accepted values. A more general form of the profit-maximizing function in a government context would be the optimal provision of a public good in comparison to the cost to provide the public good. For national security, there are three forms of a possible objective function flowing from this general form: maximize defense capability given a budget, iteratively adjust the budget until the perceived value of capability equals the cost, or minimize the budget to provide a defined level of capability. Typically, the United States has used the latter in policy—not just in logistics but overall, first setting a defense strategy and then aiming to resource to it. Shortfalls in resourcing against this strategy are then recognized and characterized as risks. As part of this process, readiness goals for equipment serviceability and on hand are set. Thus, a potential objective statement would be:

Meet readiness goals with acceptable risk at minimum total supply chain cost.[1]

In this objective statement, readiness is the constraint, with the goal of minimizing total cost subject to this constraint. The risk element includes being able to meet the readiness standard in the face of varying conditions, which should be specified. Doing so requires some slack resources or guarantees of additional capabilities or resources upon demand. Two aspects of the objective statement that have to be defined in a way that they can be operationalized are equipment serviceability requirements and the risks or range of operational demand and disruptions that the supply chain has to be prepared to handle. For spare parts, there are standard metrics and readiness goals in terms of equipment uptime that are well understood, with the services typically dividing equipment uptime goals into supply and maintenance components.[2] For operational supplies, the goals are less defined but can generally be considered as having those supplies necessary to conduct operations. Associated with both of these goals should be a minimum probability of having those supplies or meeting equipment serviceability thresholds. This can be thought of as the acceptable level of risk. An alternative objective statement would be to maximize readiness to include the potential level of risk for a specified budget.

To provide an example for operational supplies, consider water or food. The amount needed each day can be readily computed based upon the force size. The minimum amount on hand needs to be the daily demand multiplied by the interval between replenishments. If deliveries are planned every other day, then this becomes two days of supply. If the delivery interval is variable, then the minimum needs to account for the range of times. Beyond the minimum, additional supplies need to be on hand to mitigate risk. The risk would be in the form of supply lines being fully or partially disrupted. To determine the buffer level, planners should consider the range of plausible scenarios that could create disruptions and determine the potential lengths of such disruptions. Examples would be the *shamal* that hit when U.S. forces were advancing in Iraq in March 2003, stopping supply convoys for two full days,[3] and disruptions in the fuel truck supply lines through Pakistan to Afghanistan in 2010 and 2011. Judgments have to be made on which of these scenarios to protect against. This planning process would also incorporate the potential for using local supplies.

For equipment uptime, risk could be thought of similarly in terms of the lengths of supply chain disruptions that need to be handled. How to mitigate this risk could then be done in different ways. For example, this could include planning tactical inventory depth to account for some potential battlefield distribution delays, providing additional equipment to compensate for down items, having emergency resupply means in place, or preparing contingency operational plans that would account for degraded readiness.

---

[1] The new December 2011 policy instruction states, "DoD materiel management shall operate as a high-performing and agile supply chain responsive to customer requirements during peacetime and war while balancing risk and total cost," with the procedures in the manual (draft as of the writing of this report) implementing this as "The DoD Components shall structure materiel management to provide responsive, consistent, and reliable support to the warfighter during peacetime and war and establish end-to-end processes that are focused on achieving warfighter readiness goals and meeting customer needs in the most efficient way possible within the bounds of acceptable risk levels" (DoD Instruction 4140.01, 2011, and DoD Manual 4140.01, draft as of March 2012).

[2] These sub-metrics are typically defined as *not mission capable supply* and *not mission capable maintenance percentages*, reflecting attributions of downtime to waiting for parts from the supply system versus in, or waiting for, maintenance.

[3] Eric Peltz, John Halliday, Marc Robbins, and Kenneth J. Girardini, *Sustainment of Army Forces in Operation Iraqi Freedom: Battlefield Logistics and Effects on Operations*, Santa Monica, Calif.: RAND Corporation, MG-344-A, 2005.

Both concepts of the overall supply chain objective can be seen in resource allocation and decisionmaking processes in DoD today. In some services, across the board, goals for equipment serviceability are specified. This implies an objective statement of minimizing the cost to meet these goals. In other services, goals across weapon systems vary.[4] This could potentially be in accordance with varying levels needed to meet overall mission objectives. Or, more likely, it is a reaction to constrained budgets and determining the best way to allocate such budgets given the importance and role of a system and the resources required for different levels of serviceability to be achieved. Additionally, weapon system availability goals sometimes vary in accordance with where a unit is in a rotational deployment cycle. Readiness for supplies for non-weapon system maintenance support is generally not defined. But there is an understood goal of always having certain necessities available when needed, such as food, water, and toilet paper. The availability requirement in the field for these types of subsistence and personal care items is 100 percent. With regard to the acceptable level of risk, very little vulnerability or risk is regarded as permissible; the DoD supply chain is risk averse. In operations, as supply chain threats rise, buffer stocks have been increased to ensure this availability. And planning scenarios for determining war reserve requirements represent the types of risk that the supply chain has to be prepared to handle. From a readiness perspective, determining how much will be necessary for different scenarios that DoD needs to be prepared for is a more complex problem and a critical part of war reserve planning.

Guiding Principles

Guiding principles can be useful as criteria for developing policy and making decisions. They can serve as broad guidelines for supply chain design and as a check on practices, policies, and proposed changes. The first chapter of the 2004 DoD "Supply Chain Materiel Management Policy" is entitled Guiding Principles. These include a number of sound principles:

- Select providers based upon best value.
- Structure their [all DoD components'] materiel management to provide responsive, consistent, and reliable support to the war fighter during peacetime and war.
- Implement materiel management functions with standard data systems.
- Maintain materiel control and visibility of the secondary inventory down to and including retail inventories.
- Make maximum, effective use of competitive, global commercial and organic supply chain capabilities.
- Accomplish common requirements cooperatively whenever practical.
- Implement consistent structure, content, and presentation of logistics information, particularly when supporting common interfaces among the military services, Defense Agencies, and international partners.
- As early as possible in the acquisition cycle of a new program, work with the acquisition program manager and product support integrator to address logistics requirements and

[4] For example, Air Force not mission capable percentage goals vary by aircraft, with a standard goal for all ground support vehicles. U.S. Air Force Global Logistics Support Center, "AFGLSC Monthly Performance Review," May 26, 2011.

related supply chain costs (e.g., materiel, storage, transportation, etc.) within the context of total life-cycle systems management.
- Include all logistics requirements in planning and program baselines and develop them initially without any internally or externally imposed financial constraints.[5]

Based upon supply chain and business principles, RAND DoD supply chain research, and a review of policy, additional principles to consider include the following:[6]

- Decisions should reflect total cost and supply chain performance effects.
This is generally understood but should be directly articulated for a constant reminder and as part of a checklist on supply chain policies and decisions. Many problems and remaining opportunities stem from not carefully doing this. For example, it is at the heart of the first case study on theater inventory. While seemingly simple in conception, always considering total costs effectively can be difficult, requiring careful thought so as to include all of the costs affected by different options and ensuring that increased costs in some areas to reduce total costs is rewarded, not penalized. Computing all of the relevant costs can also be a challenge as well when financial systems have not been set up to support managerial accounting and decisionmaking, which is often the case—and not a problem specific to DoD.
- Where there is a scale advantage that provides a comparative advantage, it should be leveraged.
In some cases, there are supply chain design options that offer natural comparative advantages over other options, assuming effective execution of both options. These generally revolve around economies of scale that enable the use of different capabilities that offer advantages. One example is the use of a distribution center that enables suppliers to ship to DoD in quantities that will serve an area or customer base and that enable consolidated shipping to areas or customers across all of the items they need. Bypassing the DC dilutes its value. For example, consolidating shipments from an origin to a destination can enable daily full trucks between the two for next-day service at much lower cost than using a premium, next-day air shipper for each individual package. Allowing some shipments along this arc to be shipped by another mode would dilute this advantage, potentially to the point where such trucks would no longer be a cost-effective option.
- Supply chain strategies should be differentiated by supply chain characteristics and consistently applied.
In DoD, some classes of supply have been designated to be handled primarily through outsourcing the entire supply chain, including distribution. These include subsistence items and pharmaceuticals. In other cases, though, there are very similar items handled differently. For example, some hardware is direct vendor delivery (DVD) and other similar hardware is stocked in the DoD system for delivery by the Defense Transportation System (DTS). If a set of conditions suggests an optimal supply chain strategy, then it should be applied consistently. If individual analyses of items with similar supply chain characteristics lead to different solutions, then questions to raise would be whether

[5] DoD 4140.1-R, 2003.

[6] The first and third additional principles below are reflected in DoD Manual 4140.01 (draft as of March 2012) as part of the new supply chain strategies.

the options or providers considered represent effective execution and whether all of the assumptions in the analyses were appropriate.

- Outsourcing decisions should fully consider total system effects to include fixed versus marginal costs, fixed capacity constraints, and governance costs.
Similar to the total cost principle, fully understanding and determining the full cost impacts of outsourcing decisions can be difficult. Doing so, though, is crucial to making the best decisions. This is particularly problematic when there are shared resources. If outsourcing eliminates the need for use of a shared resource for part of a population but not all, the question becomes how much of the shared resource can be eliminated, if any. If reduction is not proportional, then essentially the cost for the non-outsourced items or services goes up, eliminating some of the savings that would have been projected for the outsourcing decision. This can involve distribution resources or even supply chain managers and their headquarters organizations. For example, if an item were typically shipped via scheduled truck when stocked in DLA DCs, when outsourced to direct vendor delivery (DVD), the relative shipping cost per pound of all of the items not outsourced and still shipped via scheduled trucks goes up.[7] Or for example, if a small number of items are outsourced, this may not be sufficient to reduce the number of demand and supply planners. In short, determining what costs are fixed versus truly marginal is difficult but critical. This includes any constraints on eliminating internal capacity, which can be a real concern for DoD given existing laws and the political process. Finally, including the cost of overseeing outsourcing should be part of the decision process.

A DoD Supply Chain Supply Structure or Framework

When designing and managing a supply chain, it is critical to maintain a systems perspective in order to achieve the desired performance at maximum efficiency. Given that the various supply chain processes affect each other, impacting the options available to execute other processes and the cost and performance of other processes, to maintain this systems perspective, it is critical to ensure a common understanding of the dependencies among supply chain processes, functions, organizations, nodes, and channels. For example, distribution modal options depend upon stock positioning plans, distribution modal choice depends upon stock positioning execution, stock positioning effectiveness depends upon sufficient supporting inventory, and inventory management effectiveness depends upon effective demand and supply planning and effective sourcing and supplier management.

The supply chain objective and principles lead to a framework for the DoD supply chain structure based upon the goals and the supply chain's characteristics. This framework should serve as the basis for a common understanding of the supply chain structure, the roles of each element, and dependencies to continuously factor into planning and decisions, and it should be adapted for inclusion in policy as an integration guide, such as in the form of a DoD pamphlet that augments DoD Manual 4140.01, *DoD Supply Chain Materiel Management Procedures* (draft as of March 2012). It provides general guidance on how the various DoD supply chain organizations should operate to achieve integration, in effect serving as an overview and

[7] This assumes a small portion of the shipment volume for the route is outsourced, leading to no change in the scheduled truck frequency.

primer for the DoD supply chain workforce. The more detailed required and implementing procedures in policy should align with the framework and provide more specific guidance on how organizations should execute their missions.

The following sections describe the major elements of the DoD supply chain, who is involved in each, their supply chain roles, and their critical dependencies on other supply chain elements. These critical dependencies for each element are those processes executed by other supply elements whose performance dramatically affects its cost and/or performance.

The End Users of Materiel

There are three types of end customers or functions that the DoD supply chain serves. One is the personnel who maintain and directly use equipment across DoD—the soldiers, sailors, airmen, marines, DoD civilians, and contractors in support of DoD—and those who need spare parts, tools, and packaged petroleum, oil, and lubricants (POL) to keep this equipment in working order. The second is these same personnel who need other material to execute their missions, ranging from construction material to office supplies to tools to clothing to items for personal hygiene. The third are industrial activities that make, rebuild, overhaul, upgrade, and service equipment that is provided to these personnel to execute their missions. In short, the supply chain delivers spare parts and operating supplies to field activities, construction material for new field sites and defensive capabilities, and piece parts to end item production lines.

Critical dependencies: materiel availability and wait time to receive orders placed to satisfy urgent needs.

Retail/Tactical Warehouses/Distribution Centers—The Wholesale Customers

Tactical units and other operational activities typically have a collocated warehouse to store supplies for direct use by the activity. Most are actually small DCs, which also "cross-dock" shipments for end customers, which get virtually all of their materiel through these activities. Thus, these field or retail DCs typically appear as the customer for the rest or the wholesale part of the DoD supply chain and will be referred to as the customer locations throughout this report. These activities vary in composition across the services and supported activities. They include such entities as Air Force base supply, Army SSAs, Marine Corps Supported Activities Supply System (SASSY) Management Units (SMU), Navy retail stock points in ship stores and ashore supply activities, and industrial activity SS&D sites. Some are deployable and tactical in nature; with others considered "retail" but non-tactical, such as Air Force base supply, Army SSAs for training activities, and SS&D sites. Nevertheless, the common thread among them is that they are expected to have assets on hand to support immediate needs so that the people they support can do their work and accomplish their missions as planned. They also serve as the point of entry into the supply system for unserviceable reparable items or rotable spares returned by maintenance activities. These warehouses are operated by service tactical units and civilians and contractors for non-tactical sites. The latter could be DLA or service personnel operated and managed depending upon the activity.

Their primary role is to ensure readiness or the ability to conduct operations, whatever the type. For tactical units, this can range from providing parts to immediately repair not mission capable equipment to having the food on hand to sustain personnel when it is time for a meal. In some cases, these warehouses have to be mobile and deployable, necessitating control of their size or using warehouse size as a constraint. Adding this consideration to the facts that they are large in number and serve a readiness, not cost control role, they generally count on relatively responsive replenishment to keep inventory depths low, thereby only accounting for about 12 percent of DoD serviceable inventory and 7 percent of total inventory (valuing carcasses at full price).[8] However, while they have a small percentage of the inventory value, they are critical to maintaining readiness. Stock-outs of critical parts at this level automatically lead to not mission capable equipment, and shortages of other items can prevent the conduct of operations. In multiple unpublished analyses for the Army, tactical inventory performance consistently had the greatest impact on equipment serviceability among logistics and supply chain factors.[9]

Depending upon the type of supply activity and commodity, these supply activities employ two different stockage and supply chain strategies. For predictable demand items, generally with dependent demand (e.g., each person eats three meals per day), a push strategy is employed, with the stock on hand being based upon the replenishment frequency and the amount of buffer stock necessary to cover potential supply disruptions. The level of buffer stock depends on the type of risks, such as from weather conditions or potential enemy action, in the area of operations. Some packaged POL products would fall in this category, but the primary commodities are food, water, and fuel. To illustrate, if replenishments are sent every other day, then two days of these commodities would need to be on hand. If a risk assessment indicates there is the possibility that the supply lines could be cut for up to seven days, then an additional seven days of stock might be kept unless emergency delivery capabilities could be identified.

Spare parts to maintain equipment, some class IV for expedient engineer needs (e.g., hasty obstacle), some class II, and medical items are generally managed with more of a pull strategy given either stochastic demand (e.g., most spare parts) or dependent demand based upon more variable use (e.g., obstacle emplacement). For stochastic demand items, stock levels are computed based upon desired service levels to maintain equipment and operational readiness, with replenishment triggered by hitting reorder points or minimum stock levels. For dependent demand items with variable use, stock levels can be set through scenario planning based upon operational events and the amount of materiel needed to support them.

Critical dependencies: replenishment time from the wholesale distribution and supply system.

DoD Distribution Network

The design of the distribution network consists of the number and locations of different types of DCs, how stock is positioned among them—the stock positioning strategy, and the transportation modes planned to link them and to deliver materiel to customers. The facility laydown, the location and dispersion of customers, and the conditions and infrastructure development level in a region drive transportation modal feasibility and the most cost-effective choices among the feasible options for combinations of origins and destinations. Hence, DCs and transportation options are considered jointly in this framework.

The DoD supply chain consists of thousands of geographically distributed suppliers providing most items at very low average daily rates and hundreds of field activities that need

---

[8] This is based upon Navy and Air Force working capital fund budgets, Corps Theater Automatic Data Processing Service Center data, and DLA quantity by owner (QBO) data.

[9] Peltz, Eric, and Thomas Held, "Improving Readiness for Problem Ground Fleets," unpublished RAND Corporation research, 2003; Peltz, Eric, and Aimee Bower, "The Drivers of Operational Readiness Rates: A National Training Center Analysis," unpublished RAND Corporation research, 2001; and other analyses of equipment readiness.

hundreds of thousands of different items from these different suppliers, most on an infrequent basis, resulting in low demand for most items and each supplier from each customer location. There are three important demand characteristics with regard to these customer locations with respect to distribution network design. One is that there are significant similarities in terms of the items various groups of them demand (e.g., supply activities that support squadrons of a given aircraft type), the daily aggregate demand from each location can be quite high, and multiple field warehouses are often located on the same installation or base, producing even larger daily demand for such geographic locations. This produces value in having a distribution network that consolidates shipments across items and suppliers focusing on how to best connect and deliver from these dispersed sources to relatively large customer locations that are more akin to retail outlets, such as auto parts stores, and local distributors than to household-like customers.

The DoD distribution network is relatively well conceived to efficiently handle this problem with central DC hubs or SDPs that enable consolidation across customers or field activities from a supplier perspective and consolidation across suppliers and items from a customer perspective. In other words, for a supplier location, SDPs consolidate demand across their customer bases, enabling planning for the aggregate demand across all of these customers and enabling efficient transportation to one or a small number of central hubs. Shipments from geographically proximate supplier locations could also be combined for consolidated transportation efficiency. Similarly, consolidating the daily volume for customers or geographically proximate customers across all of the items they need and all of the suppliers that provide this materiel enables efficient yet rapid consolidated transportation to them. As in any similar type of supply chain and distribution network, SDPs provide a cross-over point for the switch from economies of scale across customers for suppliers to economies of scale across suppliers with respect to distribution to customers. The overseas DCs or FDDs extend this role for bigger, heavier, higher-volume items needed in their regions, enabling replenishments by sealift using full containers composed of a mix of these items.

Additionally, the DoD system has effective practices to handle the situations in which these upstream and downstream economies of scale do not apply. When supply activities support small, isolated units and have low demand, consolidated transportation options no longer remain cost efficient. If items are needed by these types of customer locations as well as larger ones, then the item should still be stocked in the DoD system but commercial express carriers would be used for delivery to the low demand locations. Alternatively, when supplier and customer combinations support these types of economies of scale without a middle DoD DC layer, DVD and prime vendor programs may be the best choice. This can be the case for large volume, large commodities such as lumber, subsistence items, and bulk fuel. Another situation in which these programs may be the best choice is for commodities with high commercial commonality for which there are commercial distribution providers, such as for pharmaceuticals.

In addition to SDPs and FDDs, there are also forward distribution points (FDPs). These three types of distribution depots are operated and managed by the DLA in the DoD distribution network and serve distinct purposes.

SDPs

These DCs are intended to be the primary point entry of material from suppliers into the DoD distribution system. They store a wide range of material to replenish other DCs and field activities in their regions in CONUS, with two SDPs having designated roles for replenishing the

OCONUS FDDs in two different "halves" of the world. They also provide DS to designated customers in their areas of responsibility when those customers' DS supply activities or DCs do not have the items. Some SDPs also have a retail role when an SDP is collocated with an industrial activity.

Accordingly, they serve two primary roles:

- They efficiently consolidate orders to suppliers and first destination transportation (FDT) by aggregating demand for a given item and supplier location across all customers in a given region or regions.
- They efficiently consolidate second destination transportation by aggregating demand across items and supplier locations for customer locations.

When materiel is concentrated at an SDP in CONUS, it can be shipped to moderate and large-sized installations in frequent scheduled trucks, enabling short wait times at low transportation cost supporting direct readiness needs and enabling shallow depth in tactical/field supply activities. This same method can be used to replenish FDPs, keeping their inventory needs and thus total system inventory low. For OCONUS, the materiel at SDPs can be put in well-utilized containers to replenish OCONUS FDDs for items they stock. For large OCONUS locations, SDPs can build pure pallets to minimize downstream "touches" (the number of times an item has to be handled or repackaged) or accommodate more austere theaters without robust break bulk and sorting capabilities and secure ground transportation networks. For scheduled truck and pure pallet shipments from SDPs to be effective, though, most items need to be stocked at the supporting SDP of the designated customer(s). This enables high facing fill, supporting more frequent, scheduled deliveries and higher transportation utilization.

Critical dependencies: changes in operational plans that will significantly affect demand, changes in stationing plans, deployment plans, replenishments from suppliers.

FDDs

These are OCONUS DCs that store material to replenish tactical/retail warehouses in their areas of operation and that provide DS to customers in these same areas of operation for some items. They are replenished primarily from SDPs, with some replenishment from FDPs for certain types of items such as service-managed reparables. Their purpose is primarily one of cost control.

The choice between OCONUS and CONUS inventory as the first line of support or tactical/retail inventory replenishment dramatically influences the cost of responsive support for high-priority requisitions and the replenishment of overseas tactical inventory locations. There are two main ways to ship supplies overseas: by air or sea. Sealift is slow but cheap. Airlift is fast but expensive. OCONUS FDDs are a way to harness the low cost of sealift while providing responsive support to satisfy the required delivery dates (RDDs) of high-priority demands and tactical inventory replenishments. However, taking advantage of this capability requires more inventory and additional touches in the system, with materiel being issued and receipted an additional time. Thus, this distribution system option should be used only when the overseas transportation savings outweigh the additional inventory and materiel handling costs.

For items that are sometimes or always needed quickly by customers, to determine the lowest cost distribution network option choice for each item in each region, the three costs—

inventory, transportation, and materiel handling—can be computed for the two different support options to determine whether it would cost less to stock the item in an FDD or provide it to tactical/field supply activities via airlift from CONUS SDPs upon demand. In general, low-cost, high-weight, high-demand items should be stocked in FDDs, with replenishment via sealift. In the more extreme cases, the additional inventory and materiel handling costs pale in comparison to the airlift cost avoidance. A very small percentage of items dominate the demand for transportation, and when the most expensive of these are excluded, stocking these items in FDDs provides the majority of FDD value. Other items should be stocked only in CONUS to be airlifted when RDDs so demand. For responsive delivery of expensive items, airlift cost for these items can be much less than what the additional inventory for FDD stockage would be. For responsive delivery of low-demand, small items, the marginal cost of airlift is less than the additional materiel handling cost that would be incurred with FDD stockage. The value proposition for theater inventory is to spend a little more on inventory and materiel handling to save a lot on transportation, while providing good service, for select items. Conversely, the value proposition for strategic airlift of secondary items for sustainment is to spend more on transportation to save more money on inventory and materiel handling.

However, there are some theaters, particularly secure regions with well developed economies, some locations within theaters, and some specific items for which FDDs provide faster support than does airlift from CONUS. When an FDD response time advantage would reduce tactical/retail inventory requirements or improve readiness, then the combatant commands and services should work with supply chain providers to determine which items to add to the theater FDD(s) for this response time advantage. If an item is added solely for a response time advantage—i.e., stocking it in the FDD with replenishment via sealift would increase total costs—then the FDD inventory of the item should be replenished by airlift to avoid increasing inventory requirements and total costs. If sealift replenishment is cost-effective for replenishment, then an item would qualify for FDD stockage based upon cost considerations alone, as described in the prior paragraph.

Additionally, FDDs in some theaters directly support customers rather than providing support to them through tactical/retail supply activities. In effect, these FDDs are serving as a tactical or retail supply activity in addition to their broader, theater or regional role. In such cases, items to meet direct customer needs should be stocked at the FDD.

Critical dependencies: materiel availability in CONUS to support replenishments, OCONUS-positioned war reserve secondary item inventory for starter stocks in new areas of operation.

FDPs

These are warehouses that store unserviceable material for induction into collocated industrial activities, often store the products produced at these activities, and provide the material to support the production lines. FDPs serve a role in replenishing FDDs and field activities and directly support field customers primarily with respect to the items produced at the maintenance activities collocated with the FDP. When an FDP is the dominant user of an item in a region of CONUS, suppliers deliver that item directly to the FDP. Otherwise, they are replenished by SDPs.

FDPs provide rapid replenishment to line-side maintenance stocks and rapid depot maintenance support when line-side stock is not available. Thus, FDPs must stock items in support of collocated maintenance depots. SDP stockage with scheduled truck-based replenishments can be used to minimize replenishment times to FDPs and to directly support depot operations for items not stocked in FDPs. When the maintenance depot is the sole or primary source of demand for an item in a region, vendors will directly replenish the FDP, making the FDP what is called a *buyback* location.

For some expensive items, the most cost-effective DoD supply chain option is to centralize inventory at one location and use express transportation to serve all customers from there. This is the case for very expensive items that would drive very high inventory costs if inventory were distributed. These will generally be reparable items, which are often repaired at one depot maintenance activity. In these cases, the FDP becomes the most cost-effective location for the centralized inventory of the item.

Critical dependencies: replenishments from SDPs, maintenance plans.

Distribution Modal Choice for Shipments to OCONUS Customers

To fully leverage the value of FDD stock, it must be integrated with distribution planning, to include where to source customer orders from and theater transportation network planning. The first wholesale source in the source preference logic for a region should be the FDD. Theater transportation planners should establish reliable, consistent truck networks from the FDD when feasible; otherwise, they need to establish reliable theater air shipment service. If neither can be provided from a location collocated with an FDD, then FDD stock should become limited to a very small number of very transportation-intensive items, which could be delivered by truck convoys, or not used at all.

When items are not stocked in an FDD, then choices in CONUS include whether or not to stock the item in SDPs and the distribution mode for delivery. Sealift should be used when long-lead planning supports the associated distribution times, such as for major construction projects, and for other requisitions with RDDs within sealift performance standards. Service air clearance rules and Defense Transportation Regulation (DTR) policies regarding the use of sealift define these two situations. Otherwise, airlift is needed to provide sufficiently responsive delivery.

In more austere theaters, without a robust truck network with sufficient security for predictable delivery and without high capacity, high throughput break-bulk and sorting capabilities, two types of airlift-based distribution provide responsive delivery, with a third providing somewhat slower response but sometimes being the only feasible option:

- Single supply activity or single location (with the location having multiple supply activities) pallets, called *pure pallets*, with materiel issued from the supply activities' designated SDPs, enable rapid delivery with limited theater sorting and pallet building burden directly from CONUS through an en route base to a collocated airfield, security and conditions and airfield capabilities permitting, or to the nearest theater consolidation and shipping point (TCSP).
- Worldwide Express (WWX) services provided through blanket contracts with commercial shippers provide rapid delivery for smaller customer locations and items not at SDPs for items up to 300 pounds. As a theater develops, these services can often extend to most locations, but early on, delivery may have to be to a TCSP.
- Theater-level pallets sorted in theater with some attendant delays have to be used for delivery of heavy items beyond the WWX 300-pound limit for non-pure pallet locations.

Thus, these smaller demand locations would have to stock greater relative depth of these heavy, critical items than other locations.

For pure pallets to be an effective option, the supply activity or an operating base must have sufficiently large and consistent demand volume to enable short pallet hold time at the SDP consolidation and containerization point (CCP) while still achieving good pallet utilization. Additionally, the SDP must have inventory to cover a high percentage of the demands or high facing fill to meet these two criteria. Otherwise, WWX is the most cost-effective primary airlift distribution option for supply activities and their supported customers. In this case, some military-managed airlift with theater sorting will have to provide supplemental service for heavier items.

TCSPs integrate flows from different sources for delivery to customers providing resorting and consolidation by customer location and intermodal transfer capabilities. These are operated by service-provided units and agency and contract personnel.

Few customer locations have sufficient demand for well-utilized pure ocean containers. Additionally, if stock is positioned according to the design intent described here, mixed containers with sorting at TCSPs will provide sufficient responsiveness for sealift-based shipments from CONUS, which would be used for items with extended delivery date requirements. Thus, customer locations designated to receive pure containers should be limited, with most direct sealift delivered to TCSPs in mixed, well-utilized 40-foot containers, allowing for the minimization of transportation cost.

In more austere theaters, TCSPs integrate flows from FDDs and CONUS and provide intermodal transfer capability to employ theater transportation as necessary, and they provide some sorting and consolidation capability for non-pure pallets and containers.

In more mature theaters with a robust break-bulk and sortation capability and security conditions that enable reliable, truck-based scheduled delivery such as in South Korea and Europe, a third option exists: theater-level pallets with shipments for all or multiple locations in theater. These shipments are then sorted by TCSPs for delivery by truck.

Critical dependencies: stock positioning, deployment plans, restationing plans.

Distribution Modal Choice Within CONUS

Scheduled transportation service enables rapid, cost-effective delivery on routes with sufficient volume for high utilization. Within CONUS, when installation volume or the volume across installations along a route from an SDP is sufficient for well-utilized trucks at sufficient frequency to meet issue priority group (IPG) 1 time definite delivery (TDD) standards for delivery from an SDP to the installation(s), scheduled truck service should be used. Under these conditions, it provides service similar to that of next-day air providers at close to full-truck load transportation costs. Otherwise, priority differentiated service balances cost and customer needs. In such cases, depending upon item size, the amount of shipments sent to the same location from the same DC on the same day, and shipment priorities, next-day air providers, commercial air freight, small-package ground carriers, or less than truckload (LTL) service will be used.

Critical dependencies: stock positioning, deployment plans, restationing plans.

Transportation Management, Assets, and Facilities

Transportation management organizations manage and contract for the transportation assets and nodes ensuring sufficient capacity and responsiveness to customer needs while working to minimize costs. The services have commands to manage air, sea, and ground assets, with USTRANSCOM providing an oversight and integrating role. The assets they manage include trucks, container ships, and aircraft—both owned and operated by DoD and the private sector, with the latter accessed through a variety of contractual arrangements including charter, standard commercial services, and specialized contracts. DoD assets are owned by the services and operated by tactical units. Aerial and sea ports of embarkation and debarkation serve as entry and exit points for the air and sea portions of overseas shipments. Aerial en route bases can serve as hubs for the transfer of materiel from inter- to intra-theater aircraft and as well as refueling points. The aerial and sea ports and en route bases can be U.S. bases, commercial airports and seaports to which the U.S. has access, and foreign military bases for which access has been granted. They are operated by service-provided units, contract personnel, or accessed through standard commercial services. Another facility type, TCSPs, was described earlier.

Transportation management determines the best specific transportation asset options and providers to deliver and reposition materiel based upon the distribution network plan, ensuring sufficient capacity and the ability to meet delivery time needs while controlling cost. While the DC customer consolidation strategy should consider utilization of trucks, pallets, and containers, transportation managers should also monitor such metrics and coordinate changes when plan revisions would be more cost-effective. Additionally, transportation managers determine the need for ports of debarkation, ports of embarkation, and en route bases along with coordinating with DC and theater logistics managers on the need for TCSPs. Finally, they need to determine the best supply routes for new and ongoing operations, to include risk assessments and contingency planning to ensure that flows can be maintained in the face of potential route disruptions.

The management function includes setting up long-term contracts as well as expedient contracts with commercial carriers, which play a major role in the transportation of sustainment materiel. CONUS transportation, including various air and ground modes, and OCONUS sealift for sustainment materiel are provided solely via commercial carriers. OCONUS ground transportation is provided through DoD trucks, standard commercial carrier service, and specialized theater logistics support contracts, depending on the region and security conditions. OCONUS air options include standing express carrier contacts, military aircraft, military-managed chartered aircraft, and commercial air freight carriers. The use of express carriers versus the other options depends upon customer type and supply source, with the use of the others depending upon a mix of considerations to include the prices of different services, security conditions, and other needs for military aircraft.

Critical dependencies: DC shipment consolidation plans, aerial port shipment consolidation plans, deployment plans, operational plans, restationing plans, stock positioning plans.

Non-Stocked Items

Direct vendor delivery, which bypasses the use of the DoD DCs and the use of the DTS in CONUS, is a supply and distribution strategy employed by supply organizations for some items when it lowers costs and/or improves performance. Its use should be transparent to customers, with the intent to provide at least the same level of responsiveness for organically stocked items based upon whether they would be stocked at SDPs and FDDs and the likely distribution modal options. Generally, DVD of big, heavy, high-volume items to OCONUS locations should be limited unless the supplier has an OCONUS DC that would provide similarly responsive and cost-effective distribution. Otherwise, the supplier has to use excessively expensive air transportation to meet most RDDs. Depending upon the analysis, this OCONUS transportation cost issue should either lead to stockage in the DLA distribution network or dual channel support by which DVD is used in CONUS, with DLA DC stockage used for OCONUS support.

There are two natural cases for the use of DVD embodied in large-scale prime vendor programs in which the provider manages the entire supply chain, delivering directly to customers in CONUS and to designated DCs or customers OCONUS. These include larger, high-volume commodities for which volume is sufficient to achieve scale economies in distribution and transportation (e.g., full-truck deliveries to CONUS installations or full containers for OCONUS), such as subsistence items. The second case is when the commodity is commercial and commercial supply chain providers have distribution networks that can also serve DoD. The classic case is pharmaceuticals and other common medical supplies. Bulk fuel fits with both cases.

Critical dependencies: restationing plans, deployment plans, operational plans, end item maintenance production plans, and changes in retail stockage requirements.

Purchasing and Supply Organizations

These organizations manage the life cycle of the items they manage, to include sourcing, contracting, inventory management at the wholesale level, material release for sale to customers, cataloguing, and disposal. These organizations consist of the DLA supply chains, the service material and system commands, and the U.S. General Services Administration. Within their organizations, their roles can be divided into two main sets of process: materiel management and procurement and supplier management.

Materiel Management

Inventory planning and demand and supply planning need to provide sufficient inventory to accomplish two objectives. One is ensuring sufficient inventory to support the stock positioning plan to ensure deliveries can be made from the right location to support efficient distribution while meeting customer needs. The effectiveness of the distribution network planning rests on tight integration with materiel management planning and execution. The second objective is ensuring sufficient materiel availability to meet readiness needs. Thus, for consumable and reparable items, materiel managers should determine the minimum inventory needed to both maintain readiness and minimize total costs. Combining the two objectives, in some cases, minimizing total costs will require more inventory than needed to meet readiness needs. For reparable items, they also need to develop repair plans in coordination with maintenance activities.

Materiel managers can also play a key role in supplier effectiveness by collecting, synthesizing, and enabling the sharing of information valuable for planning. The suppliers can use demand and inventory data along with operational planning information that is synthesized and converted to demand forecast implications to improve their internal planning and could engage in collaborative planning with DoD. In turn, it is imperative for materiel managers to be engaged in sourcing and supplier management and to ensure those managers understand the implications of differing levels of supplier performance on overall supply chain costs and performance.

Critical dependencies: procurement or order lead times, order size constraints, retrograde processes, repair capacity and lead times, restationing plans, deployment plans, operational plans, end item maintenance production plans, changes in retail stockage requirements, and plans to introduce or phase out end items or components.

Procurement and Supplier Management

Procurement personnel select suppliers, develop contracts, and manage suppliers in coordination with supplier relationship managers. The lead times and order quantities these processes produce directly drive inventory requirements based upon the stock positioning plan and materiel availability needs. So developing effective supply strategies and working with suppliers to continually improve performance is an important element of maximizing supply chain efficiency, resulting in potentially lower demand, lower prices, and lower inventory requirements. As part of this, ensuring suppliers meet contractual requirements that reflect supply planning assumptions is critical for effective implementation of materiel management plans, which in turn are critical to stock positioning and distribution.

Many suppliers provide multiple categories of items and provide parts to sustain multiple end items, cutting across commands within services and even across agencies and services. Thus, procurement and supplier management personnel should integrate horizontally to improve efficiency of these functions as well as to improve the overall management of the supply base. This is in addition to ensuring they integrate vertically with all of the downstream elements of the supply chain.

Critical dependencies: significant new or changes in restationing plans, deployment plans, operational plans, and end item maintenance production plans that will significantly affect demand forecasts and plans to introduce or phase out end items or components.

Factories and Maintenance Facilities

These locations produce new consumable supplies or renew reparable items. They consist of commercial producers of consumables and new reparables; service-owned maintenance activities to include Army Depots, Air Logistics Centers, Fleet Readiness Centers, Naval Shipyards, and Marine Corps Logistics Bases; contract depot maintenance providers; and service field sites that provide repair capability for global or regional support.

Commercial suppliers produce and ship material in accordance with contracts and orders placed against those contracts. Their lead times drive inventory requirements, and how well they execute against promises determines whether the supply chain will meet its intended service levels and fill orders from the desired locations to minimize total costs. Similarly, internal and external repair sources remanufacture class IX reparables in accordance with material management plans to fill serviceable inventory and customer requirements. The cost of material, to include depot maintenance labor to produce serviceable class IX reparables, is the largest element of supply chain cost, so managing the costs of new materiel and the repair of reparables is critical to overall efficiency.

Critical dependencies: retail parts availability for reparable repair, on-hand inventory, changes in demand forecasts based upon changes in plans, and contract type.

The Overall Supply Chain Structure

Putting these all together creates a structure with defined, dependent roles as shown in Figure 4.1. Tactical/retail inventory enables readiness to conduct operations and ensures the parts necessary to execute depot production are on hand at the line. Responsive replenishment to these inventory locations is provided through several different means, depending upon the customer type, location, and item. For overseas locations, the responsive options are from an FDD or from CONUS, with various types of airlift-based distribution. With sealift, FDDs provide low-cost, responsive support for certain types of items in specified regions. Airlift provides the lowest cost, responsive option for other items in these regions. The best type of airlift service depends upon the region and its level of security and development, customer size and whether it is collocated with other customers, and item, with item types also affecting CONUS stock positioning and thus the choice of different airlift modes from CONUS. Commercial express service is best for low-volume customers, centrally stocked expensive items, and stock not at an SDP for high-volume customers. Pure pallets on military managed air—either charters or organic aircraft—are best for high-volume customer locations for stock at the region's supporting SDP for customers in austere theaters without robust ground transportation networks and TCSPs. When there are robust ground networks with TCSPs, a mix of theater pallets and pure pallets will often be the best option. And mixed customer pallets are needed for the shipment of large, heavy items to low-volume customers.

SDPs enable low-cost second destination transportation for customers on large-volume installations or for customers on smaller-volume installations that are on high-volume routes linking multiple installations. They also enable lower-cost order fulfillment for suppliers by allowing suppliers to send shipments to central hubs based upon aggregate regional demand. Scheduled trucks in an overseas theater or CONUS provide responsive inexpensive transportation from an FDD or SDP, respectively, leveraging the value of concentrated stock positioning at SDPs. FDPs ensure industrial activities have the parts on hand to execute planned production. Supply management organizations ensure stock is at the right place to take advantage of distribution system economies of scale, when appropriate, or concentrate inventory of expensive items when that is the best solution in order to minimize total supply chain costs, and they ensure the system has enough—and just enough—inventory to meet service-level goals and execute the stock positioning plan. This includes working with suppliers to minimize lead times and order quantities while not driving up item prices. Transportation management keeps the transportation plan synchronized with stock positioning and ensures responsive delivery upon demand, using the lowest cost options that meet transportation needs. Overall supply chain management—systems and people—keeps all of these capabilities tied together in both planning and execution. In planning, it ensures all of the dependencies are considered to produce the best overall supply chain solutions, monitoring the system to determine when plans should shift. In execution, it conducts process monitoring and control to ensure processes are being executed to standard and to plan.

This sets up the following general business rules or policies as depicted in Figure 4.2:

- Tactical/field/retail supply: stock essential "unpredictables" and continuously used consumables with dependent demand.
- Theater distribution assets and managers: maintain the capabilities for rapid deployment, setup, and continuous adaptation to meet evolving customer needs in terms of locations and volumes.
- FDDs: stock big, heavy, and high-demand items that have low costs per pound and readiness items for customers directly supported by an FDD.
- Sealift: replenishes FDDs and directly delivers long-lead-time planning items or discretionary items to overseas customers.

Figure 4.1
The Roles of the DoD Supply Chain Elements

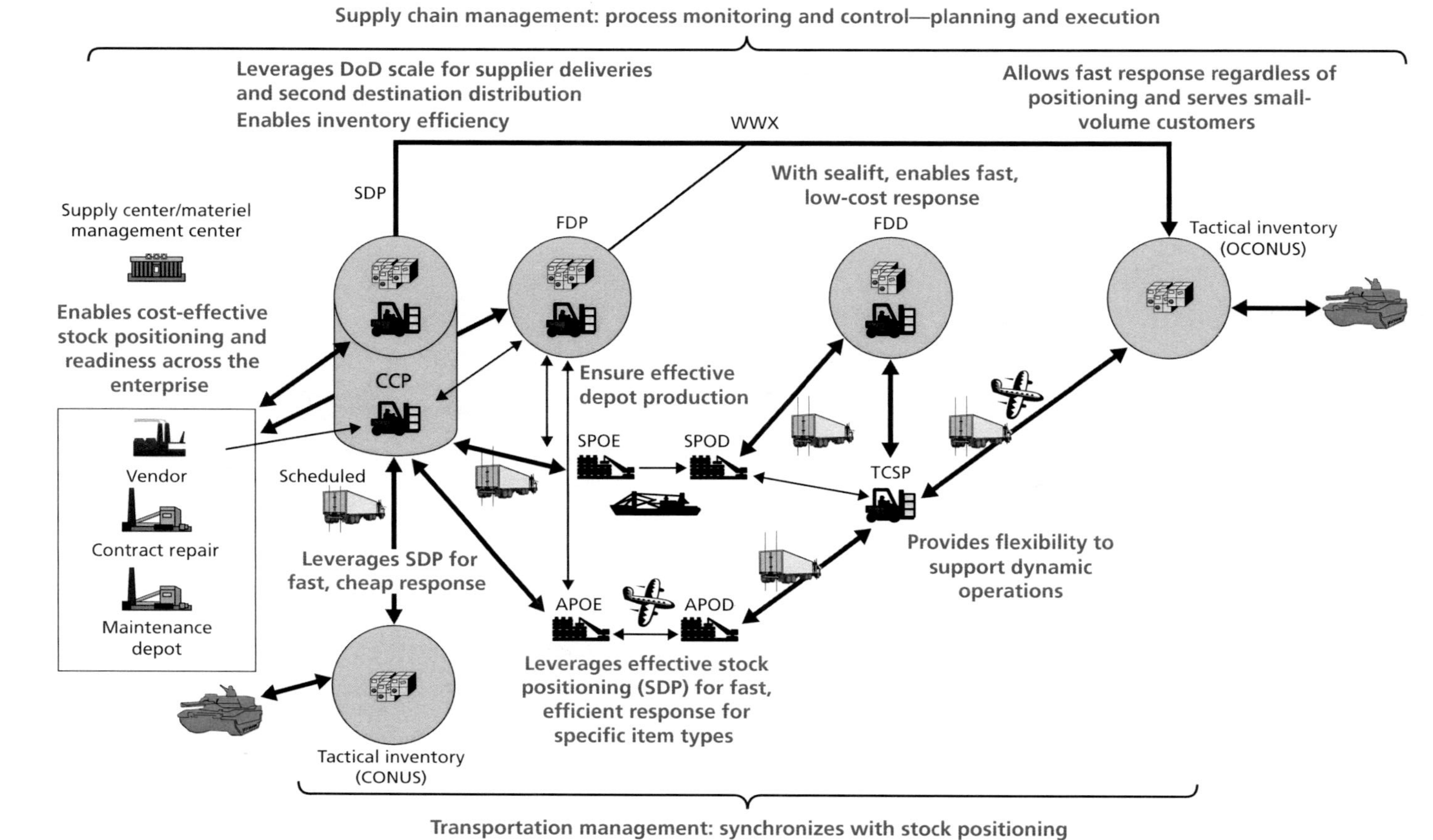

NOTE: All abbreviations can be found in the Abbreviations List.
RAND *TR1274-4.1*

Figure 4.2
Key Policies for Management of the DoD Supply Chain Elements

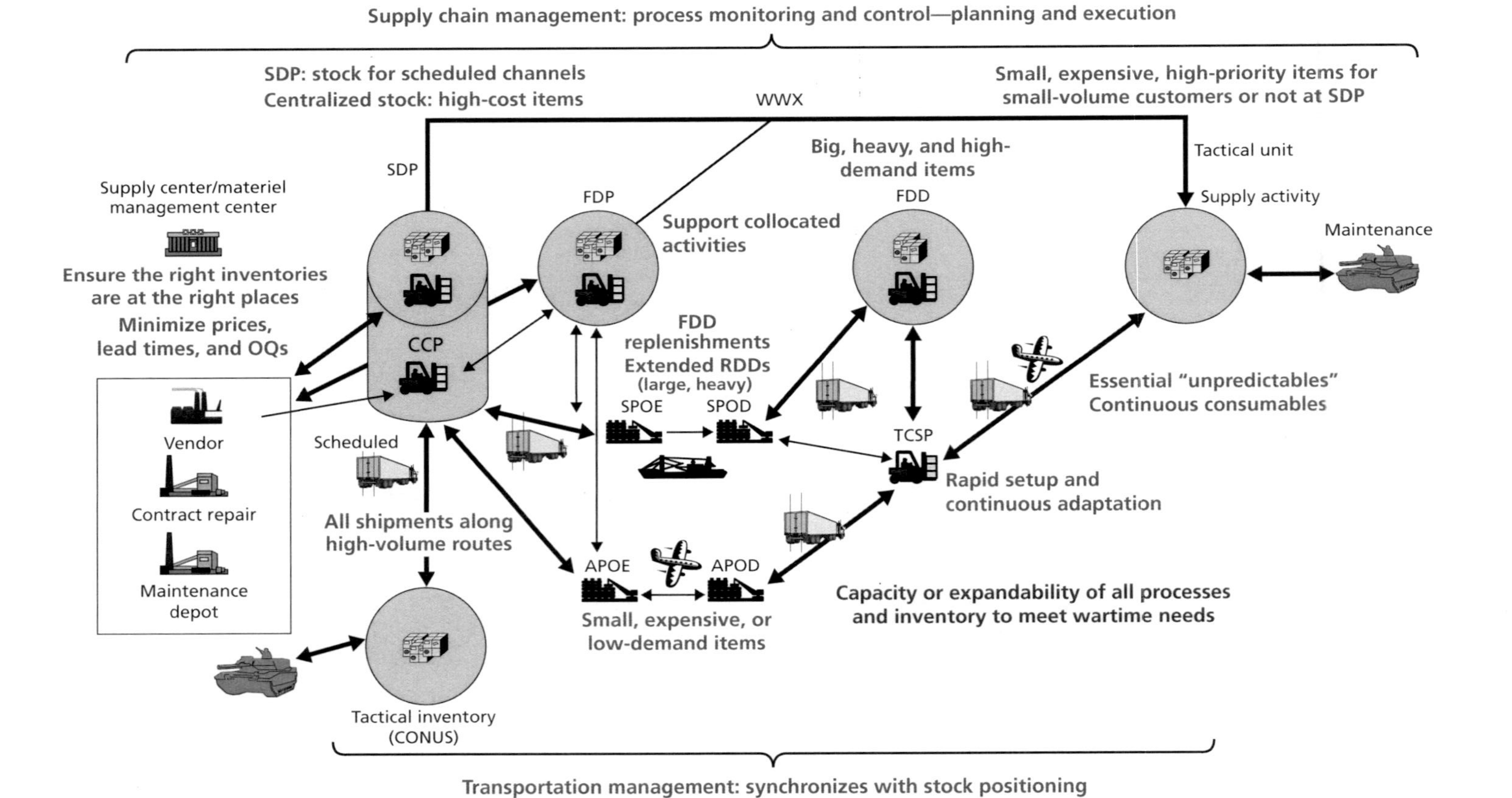

SOURCE: The figure was adapted from Eric Peltz, Marc Robbins, Kenneth J. Girardini, Rick Eden, John Halliday and Jeffrey Angers, *Sustainment of Army Forces in Operation Iraqi Freedom: Major Findings and Recommendations*, Santa Monica, Calif.: RAND Corporation, MG-342-A, 2005.

RAND *TR1274-4.2*

- Commercial small package airlift (OCONUS): used to ship small, expensive, high-priority items to low-volume customers or to fill high-priority orders for larger customers when the needed item is not at the customer's supporting SDP.
- Military-managed airlift (palletized for customers or locations): delivers smaller, expensive, or low-demand items to large-volume customers or locations when such materiel is at the customer's supporting SDP.
- Military-managed airlift (mixed cargo): used for materiel that is needed quickly and cannot be processed at DLA CCPs or that is going to low-volume customers and is too large for commercial small package airlift.
- SDPs: stock material in support of high-volume customers/locations to enable consolidated second destination transportation across items and to enable consolidated FDT.
- FDPs: stock material to support collocated maintenance activities.
- Centralized inventory at single FDPs or SDPs: very expensive items for which inventory costs dominate total supply chain costs so that having just one wholesale stockage location minimizes costs.
- Supply centers/materiel management centers: manage suppliers and plan buys and inventory to ensure sufficient stock is available of the right items at SDPs, FDPs, and FDDs to replenish tactical/retail inventories and directly serve customers in the most cost-effective manner, while also working with suppliers to minimize item prices, lead times, and order quantities.
- Overall management organizations: ensure all elements stay coordinated and provide oversight to ensure effective execution.

CHAPTER FIVE

Enabling Mechanisms

Enabling mechanisms are management and other approaches that engender execution in accordance with policy and planning intent. They include incentives to act in a way that is best for the total supply chain, metrics to understand individual process and functional effects on the total supply chain and other processes and functions, budget accounts and lines that enable and encourage people to take the best actions for the total supply chain, decision rights that create spans of control or influence that support integrated action, tools that enable people to understand the total system effects of their decisions, information systems that ensure that the requisite data for these tools are available, and career development that imbues people with the knowledge and capabilities to act in the best interests of the total supply chain both in formal planning and in ad hoc decisionmaking.

Incentives: Metrics and Budgets

A key management approach and broad category of enabling mechanism is the use of incentives. In the government, these are limited primarily to evaluations that affect promotions, awards, and the use of metrics to spur competition and influence behavior, with limited ability to use direct financial incentives. With respect to supply chain design and management, this starts with assigning responsibilities to each organization or manager and having the means to assess performance with respect to those responsibilities. If these responsibilities and associated metrics are not fully aligned with intended policy (assuming it reflects supply chain integration), then gaps in supply chain integration may occur.

Virtually all supply chain functions and processes affect others. If these interaction effects are not monitored through metrics used for feedback and accountability, then the way the processes and functions are executed may not be aligned with the overall supply chain intent and effective integration. Such gaps appear to have been a significant contributor to DoD supply chain "siloization," in which organizations optimize their own functional responsibilities, tied to narrow metrics and incentives, at the expense of other processes and the total supply chain. In short, there are a number of cases in which people and organizations in the DoD supply chain have been affecting downstream processes without feedback via metrics and without accountability and responsibility for these effects. These cases tend to also be associated with people and organizations driving downstream costs they do not have to "pay," because the costs are not in their budgets. Their actions, instead, affect the budgets of other organizations.

In the realm of inventory, currently in DoD procurement, lead times and order quantities drive inventory levels. Supply organization planners tend to be responsible for inventory costs and benefits—inventory value and materiel availability. However, given the predominant influence of lead times and order quantities in determining inventory requirements, contracting and supplier management personnel and organizations play a significant, if not the most important, role in determining inventory levels. Yet, traditionally, they have not been responsible for inventory-related cost and performance metrics or been given feedback on their effects on inventory. The focus instead tends to be on item prices and whether suppliers meet performance requirements, even if the requirement is a relatively long lead time.

As described in the first case study, full DLA implementation of optimal FDD stockage with safety levels incorporated into wholesale inventory planning took several years. During this time, supply organizations did not have responsibility for their effects on stock positioning and did not see metrics showing the effectiveness of forward stockage. While they were having a major influence on OCONUS airlift costs through insufficient inventory replenishments to FDDs, they were not made aware of these costs and did not have metrics to track them. These costs were not in their budgets but paid for by the services that needed the airlift shipments due to insufficient OCONUS stockage. As a result, the supply organizations were reluctant to fully implement the cost-minimizing theater stockage concept because the additional inventory cost would run against their metrics and goals, and conversely they would not get credit for or budget benefit from the reduced transportation costs. This dual combination of barriers, such as budget lines and metrics that could show degradation as the result of increased inventory for OCONUS stockage, and lack of facilitators that would encourage or aid improved OCONUS stockage made change doubly hard.

Similar to the OCONUS stockage situation, today's supply chains are not "graded" on CONUS stock positioning, which impacts the opportunity to use scheduled trucks and thus second destination costs. Again, they do not directly see the transportation cost impacts of stockage level and positioning decisions. In part, this has led to deferred procurement decisions when one location is out of stock, falling back on filling orders from the "wrong," less cost-effective location.[1]

Also in CONUS, the ability for local managers to opt out of scheduled truck routes is reinforced by the lack of metrics showing them the cost impacts and the fact that, while they are affecting total DoD costs, their decisions to opt out are not affecting what they are charged for distribution service. Instead, they are affecting DLA's budget and the overall surcharge and hence what all customers in the system pay, but not their own budget or metrics, when they choose to use premium transportation in lieu of equally capable scheduled trucks. Transfer pricing with differential charging, foreshadowing another enabling mechanism—financial methods—would be a way to signal inefficient choice of distribution options.

With regard to scheduled trucks, the route structure has not been kept optimized, leading to low utilization and thus excessive transportation cost. Distribution planners have not been directly responsible for and measured on second destination cost, which goes into supply chain costs and thus surcharges. Nor have they had metrics showing truck utilization.

Similar to the scheduled truck route planning, earlier in the course of OIF and Operation Enduring Freedom, container and pallet route plans were not kept up to date and optimized on a systematic basis, leading to higher airlift and sealift costs. For example, volume for a destina-

---

[1] In 2011, a DLA Stock Transport Order Integrated Process Team developed logic to determine when it is economical to reposition stock versus ship from out of area distribution centers or buy more stock for delivery at the needed location, and DLA implemented this new logic in February 2012, eliminating this problem.

tion sometimes declined below the level for which containers or pallets consolidated solely to serve that location would be efficient but the consolidation plan did not change immediately. The route planners did not have utilization metrics or directly face the increased transportation costs.

When USTRANSCOM started tracking utilization and identifying opportunities to increase it, they produced about $150 billion in annual savings (as of 2011) through improved aircraft utilization and about $130 million via improved container utilization.[2] In this same vein, DLA's subsistence supply chain did not have responsibility for or the budget for the transportation of food to overseas operations. Thus, its prime vendor food contracts did not have container utilization as a metric. When USTRANSCOM identified the cost impacts, DLA was able to modify incentives and increase container utilization, driving down transportation costs.

These examples all have the interactions between inventory, materiel handling and shipment preparation, and transportation in common, with procurement and supplier management interactions with these processes also coming into play in some of the situations. This is consistent with the lack of guidance on integration between these functions in policy. There has not been a policy emphasis, explanations of what the dependencies are, clear guidance on how to account for the interactions, and accordingly a lack of metrics that show cross-functional effects. Increasingly these types of metrics are being added as these problems are being understood. While budget lines still tend to be split with control of costs in one organization but the budget for those costs in another, there is some movement toward the creation of cost metrics showing people their cost effects on downstream processes in order to give them feedback and hold them accountable. A notable example is OCONUS stock positioning for which DLA is developing both process performance and cost metrics, with the latter showing both savings achieved from additional inventory investment and savings forgone from stockage shortfalls.

In short, metrics have reinforced a policy emphasis on meeting customer needs and minimizing inventory with much less attention to meeting needs efficiently and total cost. Meeting customer needs has been defined in terms of logistics response time or customer wait time and materiel availability or fill rate. Stock positioning has not been measured, which is critical for achieving required customer wait time requirements efficiently from a transportation standpoint. Cost metrics have focused on inventory, with less attention on total supply chain cost including transportation.

Critical additions to DoD metrics at the enterprise and organizational levels include stock positioning metrics to help limit use of expedited transportation to when it is the most cost-effective option, utilization metrics to drive supply and distribution providers to use transportation efficiently, and the effects of procurement lead times and order quantities on inventory, to include the generation of retention stock as described in Chapter Six.[3] DoD has been emphasizing perfect fulfillment, but the way it has been measured focuses on customers only,

---

[2] U.S. Transportation Command, USTRANSCOM Point Paper "Distribution Process Owner Strategic Opportunities (DSO)—Strategic Air Optimization (SAO) Opportunities," June 2011b. U.S. Transportation Command, USTRANSCOM Point Paper "Distribution Process Owner Strategic Opportunities (DSO)—Strategic Surface Optimization (SSO) Opportunities," June 2011c. Note that these savings include substantial materiel other than class II, IIIP, IV, and IX, including non-sustainment materiel.

[3] As of the writing of this report, the recommended metrics from the Deputy of the Assistant Secretary of Defense for Supply Chain Integration supply chain metrics group do include procurement lead time.

not on whether orders are filled perfectly from a supply chain standpoint. It does not include whether something is delivered from the optimal place. A filled order may meet the customer's requirements perfectly, but it does not meet DoD needs perfectly if it comes from a location that drives up total costs to meet the customer's needs.

Also, the way that DoD defines orders makes perfect order fulfillment much less valuable than in the private sector and redundant with wait time metrics. In DoD, all orders are at the "line" level for one item. If a maintenance work order needs multiple parts, they are all considered different orders. So too would be orders for the different types of POL needed to support a unit or the different products needed for a construction job. A more typical and useful form of perfect order fulfillment is tracking orders at the task or work order level or something like the store or supplier level (e.g., all lines ordered from a supplier to replenish a DC on a given day). Unless DoD information systems change to enable perfect order fulfillment at this higher-level notion of an order, its continued use will not be value added.

Decision Authority and/or Rights

From a supply chain design standpoint, when resources can be shared for a scale advantage, then it becomes valuable to coordinate their use. This requires coordinating the use of the potentially shared resources among all of the potential users. To achieve this, one of the principles for the design of enabling mechanisms is aligning decision authority or rights with the organization with the best systems perspective of a process so that the span of control or influence enables effective utilization of shared resources. In other words, this means ensuring that the organizations that can look across entities to make optimal systems decisions are in a position to act on these cross-organizational views and analyses. For example, for scheduled trucks to be effective and efficient in servicing a route, they need sufficient, consistent volume. To achieve the maximum potential requires that all customers on a route be served by the truck. In many cases, still, a single installation will not have enough volume to achieve high utilization with high-frequency (e.g., daily) trucks. But the distribution manager might see that two installations, with one directly on the route to the one farther away, would have sufficient volume for daily, full trucks. However, if each customer organization or supply activity on these installations can determine its mode of shipment, any that opt out, requiring high-priority shipments be sent via air, will reduce the scheduled truck's utilization, and if sufficient numbers opt out, then the truck route will not be feasible from a scheduled truck basis. Additionally, all of the cost to serve those that opt out translates directly into an unnecessary cost increase for the system. A similar problem would occur if the decision to opt out were at the installation level. In short, once volume is sufficient for a scheduled truck, any additions to its volume until maximum effective capacity utilization is achieved are free from the standpoint of the truck and reduce costs by eliminating other modes of transportation between the same origin and destination.

The only way to optimize all of the shipments from a DC is to plan the intended transportation modes considering all of the shipments in aggregate. Thus, a planner at the DC level would be positioned to develop the transportation plan. At a higher level, if some customers could be served by multiple DCs given similar distances and demand profiles, then the best DC to use could depend upon the potential for effective multi-installation scheduled truck route designs from each of the different DCs and factors such as DC workload. From this perspective, only the distribution system manager, overseeing the full set of DCs, would be positioned to design the optimal distribution and transportation plan encompassing which sources to use and how to consolidate transportation for different destinations. So for this example of second destination transportation, the organization that plans the transportation routes for the distribution system as a whole would be best positioned to develop an optimal plan. Therefore, this organization should have the planning responsibility and decision rights on distribution options, with the requirement to determine the most efficient set of options that meets customer requirements, with oversight organizations protecting the customers and ensuring that customer needs are indeed met. In general, for second destination transportation planning, creating a systems view requires planning across items, customers, and installations. So the providers, not the customers, should have the decision rights. But with decision rights should come metrics-based accountability for meeting customer requirements.

FDT is similar in that each producer or supplier location may or may not have sufficient volume to use full trucks. But production locations serving the same DCs could have shipments pooled to generate efficiencies. The organization with the view of all inbound freight to a DC or the full set of DCs would thus have the best systems view for planning and decisions on modes.

For DoD, many suppliers provide different items to support different end items or other activities. These various end items may be managed by different people, different organizations with services and agencies, and even different services and agencies. If each manages the relationship in a different way and contracts differently, then this is likely to make operations less efficient for the supplier as well as for internal DoD organizations. Additionally, there may be duplication in supplier management. In this case, the shared resource is the supplier, suggesting the value of an integrative management approach across all of the users of a supplier. If some of the users have vastly different needs, then their portion of the relationship could be managed separately.

With regard to wholesale inventory management, the systems view comes into play first at the end item application level, where it takes a mix of all of the component parts to enable readiness. So inventory levels for these items need to be planned together. When inventory budgets have to be allocated across end items, then this requires a cross–end item view. Both of these system-level views are taken into account today in wholesale inventory planning, which sets safety levels across items based upon the most efficient way to achieve an aggregate goal. Storage allocation would also need to be taken into account when it is a shared resource for the support of multiple end items, as can be the case in tactical warehouses and has even been the case in DLA FDPs and SDPs.

Additionally, given the interaction between inventory and transportation, an organization that can plan and account for both inventory and transportation needs to have responsibility for planning both, or these considerations need to be incorporated into planning in an automated way. As an example of this latter approach, DLA OCONUS DC stock positioning planning incorporates inventory, materiel handling, and transportation costs in a planning model to determine stockage breadth and depth. This in turn produces safety levels that DLA's overall inventory planning system incorporates into overall supply planning, enforcing automatic consideration of the systems view.

For retail supply, planning also has to account for the mix of items needed to maintain readiness at the end item level. In some cases, though, there is a fixed amount of storage space—a form of a shared resource—that has to be managed across end item and other operational needs. Typically, inventory planning methods take all of these factors into account, enabling a systems view.

Table 5.1 summarizes the ideal systems view for different supply chain processes, with the color coding indicating the degree of alignment between DoD decision rights and the recommended views for each process. For FDT, each supplier acts independently, with no coordination among them. For second destination transportation, the system is set up for coordination across customers and installations, and there is some integration among them by distribution system planners. However, customers can independently decide what transportation modes should be used to support them. For supplier management, some services and agencies have set up integrated relationship management across items. However, this is generally limited to within services and agencies and does not cover a wide range of suppliers. Wholesale supply and retail supply are planned by the services and agencies in ways that consider appropriate populations of items for determining how to invest inventory funding. However, a remaining weakness in wholesale supply is lack of comprehensive inclusion of stock positioning considerations.

Table 5.1
Aligning Decision Rights with the Systems Views for Process Integration

Process	Best Systems View for Integration
First destination transportation	Across items and suppliers along a route to a DC
Second destination transportation	Across items and customers Across customers for a destination or route
Supplier management	Across all items and supply organizations for a supplier
Wholesale supply	Across items for an end item Across items for other readiness applications Incorporate stock positioning
Retail supply	Across end items for a unit

Financial Controls, Methods, and Budgets

Financial mechanisms are employed to ensure effective stewardship of funding as well as to encourage efficient behavior. It is critical that they do the former in a way that does not impede supply efficiency and the design of the latter must be careful to not encourage the wrong behaviors.

A financial enabling mechanism is the use of working capital funds (WCFs) that enable dissociation of orders and payments enabling smooth inventory management of longer-lead items across FYs. Allocating WCF obligation authority (OA) ensures people do not place their organization on the hook for more money than will be available. But processes need to be in place to readily reallocate OA to ensure funding is used as efficiently as possible as new information becomes available.

The WCFs are also designed to encourage the right behavior through pricing signals developed through points of sale and transfer prices. One area where the signals do not appear congruent with efficient use of resources is returns of serviceable items. Very low credit from DLA leads to services retaining stock for reissue within the service, creating "shadow" or redundant distribution capabilities and impeding integrated demand and inventory planning.

Transfer pricing or charging mechanisms can also be used to drive desired behaviors. For example, today storage costs are charged to supply organizations based upon the amount and type of space they use. As more space is used for a given item, costs go up linearly. To encourage improved inventory management, penalties could be introduced for excessively low turns or excessive use of space, which could be defined. In other words, charges could increase nonlinearly or at certain steps. Or more could be charged for "prime real estate" in high-volume DCs that offer scheduled truck service and less for very infrequently demanded items placed in specialized low-demand DCs. This could also be thought of as introducing an opportunity cost charge when valuable space starts getting constrained in the system, particularly when that space is not being used to its maximum potential. For example, if a DC with scheduled truck service is full, preventing stockage of heavy or high-demand items provided to the scheduled truck customers, there is a transportation cost increase associated with not being able to stock such items at that location. In particular, if it is full due to items being in long supply or from storing items not demanded by the DC's scheduled truck customers, then the opportunity to gain value from shipment consolidation has been lost.

Budgets were discussed in tandem with metrics, because they tend to go hand in hand. The key message is that for an organization to act in the best interest of the overall supply chain, budget lines should give an organization the full set of options that should be considered in tradeoffs. Or, if the budget lines are narrower, organizations need clear feedback and responsibility for negative budget impacts they impose on others. A clear example of this is that supply organizations tend to drive the mode of transportation for shipments to customers through stock positioning. But these organizations have budgets for buying items and inventory only in accordance with meeting materiel availability goals. Their budgets do not include the transportation to deliver the items they manage or allow for investments in inventory to achieve stock positioning plans, so they cannot trade off inventory and transportation costs internally.

Information Sharing

Changes constantly occur in supply chain processes that affect others in the supply chain. Each such change can be encoded in information shared with other functions and organizations in the supply chain. If their processes are configured to utilize and act on the right types of information, then the entire supply chain can adjust in concert. Downstream process and planning changes become proactive rather reactive, avoiding inefficiencies that result from delayed—or no—use of information. Thus, understanding what information each part of the supply chain needs from the rest and developing processes to use this information is important to maximizing the potential of supply chain integration.

Decision Support Tools

The supply chain is a complex system, with a number of interactions and tradeoffs coming into play in a broad range of decisions. Without standard tools, it is not difficult to overlook one or more considerations. And more problematic, pulling together the requisite data and developing the methods to make integrated decisions can be very difficult and time consuming. If such data and capabilities are not available when questions arise, compiling the data (if the needed data are even tracked and archived) and developing a new capability in time to support a decision timeline may not be possible or may force compromises in the approach. Thus, ensuring the availability of standard decision support tools that take care of these analytic needs is important for broad-based consideration of total supply chain costs and integration. This includes ensuring the requisite data are generated, captured, and archived. In particular, an increased focus on total supply chain costs in decisionmaking would be facilitated by improved collection of cost data associated with all supply chain and logistics processes along with enhanced capabilities to allocate costs.

Workforce Knowledge

Policy and automation will never be able to take care of every possible situation involving decisions that impact supply chain integration; nor will they be foolproof. The better the workforce understands what supply chain integration means and the underlying concepts for how the supply chain works, the better they will be positioned to intervene when necessary or propose integrative solutions in new situations. The total workforce becomes the last line of defense. Additionally, strong supply chain integration knowledge and reinforcement of the need for systems thinking can help ensure that organizational and functional pressures do not crowd out broader considerations.

Principles for Enabling Mechanisms

Several principles can be distilled from this discussion of enabling mechanisms. These principles can be used to design enabling mechanisms, assess the root causes of supply chain integration problems, and directly assess enabling mechanisms:

- Organizational metrics and accountabilities should reflect the full set of outcomes and costs that they significantly influence.
- Decision authority and/or responsibilities should align with the organization with the best systems perspective.[4]
- Budget boundaries should align with the organizations that drive the use of resources.
- Financial controls should ensure effective stewardship of resources without inhibiting supply chain performance.
- Requisite information from other organizations should flow freely, with planning systems designed to automatically leverage the relevant information.[5]
- Decision support tools must take into account all supply chain costs and should be available for typical decisions.

---

[4] This principle has been incorporated into DoD Manual 4140.01 (draft as of March 2012) under business practices.

[5] This and the next principle are largely reflected in DoD Manual 4140.01 (draft as of March 2012) under procedures for supporting technologies.

CHAPTER SIX

Supplier and Inventory Management Integration

This and the next four chapters explore opportunities for improved DoD supply chain efficiency through improved integration. They involve integrating across functions based upon interactions or dependencies as well as taking more of an integrated systems view in performing a process, and they build upon the supply chain design and enabling mechanism guiding principles. The first focuses on the impact of supplier performance and management on inventory efficiency. The second revolves around the interactions between shipment consolidation in DCs and the impact on transportation efficiency, along with the power of taking a systems view across delivering materiel to all customers. The third focuses on the integration of supplier and transportation management, again with potential for taking a systems view across shipping materiel from suppliers to DoD. The fourth discusses taking a holistic view of positioning and repositioning materiel in the system based upon joint consideration of all supply chain costs. The fifth involves integrating financial policy with distribution system design and inventory planning along with integrating inventory management across organizations.

Where Is the Money?

Frequently termed the Willie Sutton rule, the oft-cited first step in identifying opportunities to save money is identifying where the money is. So if DoD wants to save money in providing supplies to its personnel to conduct operations and maintain equipment, the first step is determining what drives the costs. To do this, we turn to the service and DLA WCF budgets, dividing the budgets into four main categories: cost of material, people, transportation, and other. As indicated in Table 6.1, the cost of material dominates total expenses at $40 billion ($31 billion without subsistence and medical items) out of $49 billion.[1] The cost of personnel is second at $3.3 billion. These primarily comprise the people to manage the supply chains from demand and supply planning to supplier management and contracting, the people who operate physical distribution facilities—primarily the DCs, and the people who manage the enterprise. Other costs include other purchased services, utilities, travel, material for opera-

[1] Taking out "subsistence" and "medical" leaves primarily the classes of supply that are the focus of this report but this cannot be done for expenses, so both totals are provided. Additionally, because of their associated prime vendor programs, there is very little DoD inventory of subsistence and medical supplies. Note also that from a supply chain and supply management account standpoint, the cost of materiel includes the cost of depot labor to repair class IX reparable items.

Table 6.1
Working Capital Fund Supply Budgets (in millions of FY 2010 dollars)

	DLA	Air Force	Army	Navy	Marine Corps	Total
Cost of material sold from inventory	19,911[a]	6,223	9,393	4,409	144	40,079[a]
Cost of people	2,076	358	269	575	2	3,280
Other operating expenses	1,424	873	495	356	2	3,149
Other purchases from revolving funds	393	271	351	262	9	1,286
Goods transportation	495	83	109	119	—	806

NOTE: The Air Force cost of material sold includes cost of repair for non-secondary item maintenance. This was included because in its reporting, it consolidates the Maintenance and Supply Division operating expenses, precluding separate breakouts of these expenses.

[a] The DLA cost of material sold from inventory without subsistence and medical would be $11,032 million, and the total would be $31,200 million. However, the other costs associated with these two supply chains are not broken out in the WCF budgets.

tions, equipment, rent, advisory services, purchases from other revolving funds, and printing and reproduction. The final category is transportation of the material.[2]

The cost to hold inventory is not a separate category in the supply budgets but rather is embedded within several: the cost of material, the cost of people, and other operating expenses. It is typically considered to consist of the cost of inventory losses whether from shrinkage, obsolescence, or disposals of unneeded inventory due to forecast error, storage cost, and the opportunity cost of capital. We estimate the elements that produce direct annual DoD expenses from buying and holding inventory. From 2005 through 2011, disposals or reutilization of serviceable assets outside of DoD have averaged $2.5 billion worth of materiel per year. Over the same time period, disposals of unserviceable but potentially still economically reparable items averaged $5 billion of materiel per year.[3] These disposals represent material purchased but not needed and thus are a component of the cost of material in the supply budgets. Storage costs—which are embedded within two budget categories, the costs of people and other operating expenses associated with running DCs—were estimated at about $0.3 billion in FY 2010.[4] Depending upon how the unserviceable reparables are valued, this produces $4.5 billion to $7.8 billion per year of direct expenses. Above these direct expenses would be the opportunity costs of the capital tied up by purchasing the inventory and using the DoD-owned land occupied by the storage facilities.[5] Regardless, even at the lowest end of the estimate without

---

[2] Department of the Air Force, *United States Air Force Working Capital Fund (Appropriation: 4930), Fiscal Year (FY) 2012 Budget Estimates*, February 2011. Department of the Army, Army Working Capital Fund Fiscal Year (FY) 2012 President's Budget, February 2011. Department of Defense, *Defense Working Capital Fund, Defense-Wide Fiscal Year (FY) 2012 Budget Estimates Operating and Capital Budgets*, February 2011. Department of the Navy, *Fiscal Year (FY) 2012 Budget Estimates: Justification of Estimates Navy Working Capital Fund*, February 2011.

[3] Both figures are based upon Strategic Distribution Database data, with unserviceable but economically reparable items consisting of disposal of condition code *F* items.

[4] Department of Defense, 2011.

[5] Given interest and inflation rates as of 2012, the opportunity cost of capital for internal government investments is considered to be quite low. Typically, items are held in DoD inventory two to three years. For 2012, this leads to a 0.0 per-

including opportunity costs, annual costs associated with holding inventory would be the second highest individual cost category after the cost of materiel sold to customers if accounted for separately.

So to dramatically reduce costs, it is critical to attack the costs paid for material and inventory. What drives these costs? The purchase of material to meet demand is driven at the top level by two components: demand and the prices paid for items. Demand comes from two factors: operations—the nature and level of activities that need to be conducted—and efficiency of consumption, which among the classes of supply considered in this project is driven primarily by reliability and durability.

The Drivers of Inventory Costs

The primary drivers of inventory are order quantities, procurement lead times including administrative order processes and supplier delivery or production time, and safety stock, with the first two categories dominating DoD inventory requirements and position. While on-hand inventory is typically assumed to be safety stock plus half the order quantity (OQ) (and in DoD's case, expected on hand also includes war reserve, the repair cycle pipeline for reparables, stock to protect against the loss of diminishing manufacturing sources, and insurance stock for very low-demand items), with procurement lead-time driven inventory typically considered in the "delivery pipeline"—either on order or on contract, for DoD the on-hand level is significantly greater than the safety stock plus half the OQ and the unique DoD requirements. For example, in a September 2011 inventory stratification report, the DLA safety level requirement was $1.2 billion, the aggregate value of the order quantities was $2.5 billion, $0.1 billion was needed to cover diminishing manufacturing sources, $0.1 billion covered war reserve needs, and the insurance stockage objective was $1.1 billion. This produces an expected on-hand inventory of $3.8 billion. However, according to the same report, DLA actually had $12.9 billion on hand, much above even what the total would be with the lead-time demand quantity, including a significant portion considered economic and contingency retention stocks (i.e., above computed inventory needs but considered economical to keep or of potential use in future contingencies) and just $0.3 million considered as potential reutilization stock beyond these retention levels.[6] Similarly, a DoD-wide inventory stratification report for September 2009 suggests an on-hand "should-be" value of $42.1 billion with $97.8 billion on hand.[7] This DoD-level should-be on hand also includes the repair cycle requirement, which accounts for the repair time for service-managed reparables, and a substantial portion of the actual stock

cent real discount rate, down from 0.9 percent in 2010 (and a high of 6.1 percent in 1982), per Office of Management and Budget, *Guidelines and Discount Rates for Benefit-Cost Analysis of Federal Programs*, Circular No. A-94, October 29, 1992 (Appendix C, Revised December 2011). Also see Office of Management and Budget, "Table of Past Years Discount Rates from Appendix C of OMB Circular No. A-94," November 16, 2011.

[6] We note that prior to the start of operations in Afghanistan and Iraq, which increased DoD demands and thus the inventory requirement, the on hand stood at about $63 billion (DoD supply system inventory report, 2011). With the end of operations in Iraq, as the drawdown in Afghanistan progresses, and as the force size decreases, demands will decline, reducing the inventory requirement, and allowing for some "natural" drawdown of on hand.

[7] This excludes AMCOM and CECOM inventories, which were not included in the stratification report because of an information system transition. Otherwise, it includes all DoD inventory, whether held in DLA distribution centers or other locations.

on hand is reparable items, which stay in the system until they are "washed out" as no longer economical to repair or are deemed excess.

As described for the DLA-only inventory, most of this DoD "excess" on hand is economical to keep—that is, demands continue, and it will cost less to hold it in storage and draw it down over time rather than to dispose of it and have to repurchase material—or it is considered of potential value for possible contingency operations. Periodically, the services and DLA determine potential reutilization stock above these levels and generally dispose of that. So the idea is not to point out the gap between what should be on hand and what actually is on hand to suggest disposing of inventory. Rather, the intent is to understand what has driven the development of this gap and identify how to prevent more than the should-be on-hand level from accumulating for new items or current items at expected levels.

This greater than expected on hand comes from three factors, with only the last two applying to DLA given its management of consumables and all three applying to DLA and the services: forecast error accumulated over time; the fact that much of the inventory is in reparable items; and, for low-demand items, the lead-time inventory is often on hand. Forecast error is often thought about in terms of whether the forecasting method is working well. However, as will be discussed in more depth later, most of the DoD forecast error is driven by long-lead times, not the forecast methods or use of information in the development of forecasts.[8] When the demand trend shifts between when an order is placed and when it comes in,[9] if the forecast was too high, this increases on hand above intended levels and leads to excess inventory retained for economic and contingency reasons as well as the disposals when the excess exceeds what makes economic or operational sense to retain for these two reasons. Large order quantities, often used in DoD, compound this effect as they increase the amount of potential excess when demand diverges from the lead-time forecast. In effect, the longer the lead time and the larger the OQ, the greater the exposure to the risk of too much inventory. Additionally, lead times are the primary driver of safety stock levels, which are based upon lead-time demand variability and desired service levels. So inventory is driven primarily by order quantities and lead times. It is also important to note that the lead-time problem produces another impact: more exposure to the risks associated with an unexpected demand trend increase or shock. If such events drain the system of assets, it takes a full lead time to recover and begin serving customers again.

There is a common denominator for item prices, reliability, order quantities, and lead times: suppliers. How DoD works with suppliers and how suppliers perform affect these four factors. This leads to supplier performance, management, and integration with DoD as the biggest potential opportunity to cut costs. Improved supplier performance and integration could have a four-fold effect on costs:

- Shorter lead times
  - Less inventory overall
  - Less retention stock and excess to be disposed of for inventory losses
- Lower ordering costs for lower economic order quantities and thus less inventory

---

[8] Various DoD internal, unpublished studies have analyzed a wide range of forecasting methodologies, generally finding little opportunity from adopting different algorithms or approaches.

[9] This is a case in which the mean demand changes or there is a non-stationary mean over time, rather than being an issue of variance around the mean, which is accounted for through safety stock.

- Lower item prices
- Better reliability for lower demand.

And a fifth benefit is improved customer support and readiness due to shorter stock-out periods after demand "shocks." While shorter lead times translate to lower safety stock for the same service levels, which reduces cost, the nature of stock-outs would also change with shorter lead times, with the extent of problem periods getting shorter. More importantly, if there are major unanticipated shifts in demand—either higher or lower—the system will react faster, reducing customer support problems or the buildup of excess, respectively. With demand shocks or shifts, the system cannot recover or respond until a lead time away.

Through interviews, participation in DoD meetings, reviews of service and DLA metrics, DoD improvement plans, and GAO reports, it is clear that there is significant emphasis on reducing inventory. However, much of the emphasis is on the forecasting process instead of lead times and order quantities.

DoD Lead Times and Order Quantities

Cheaper, consumable items tend to have shorter lead times than more expensive or reparable items. These are primarily managed by DLA, so DLA managed items, in aggregate, have shorter lead times than those managed by the services. However, since procurement lead times are a more important factor for consumables than reparables with respect to inventory levels, we will focus the lead-time and OQ discussion on consumables. To illustrate the lead times that supply planning has to account for, Figure 6.1 shows the distribution of lead times for

Figure 6.1
Distribution of DLA Lead Times for the High-Demand (in Dollar-Term) Items

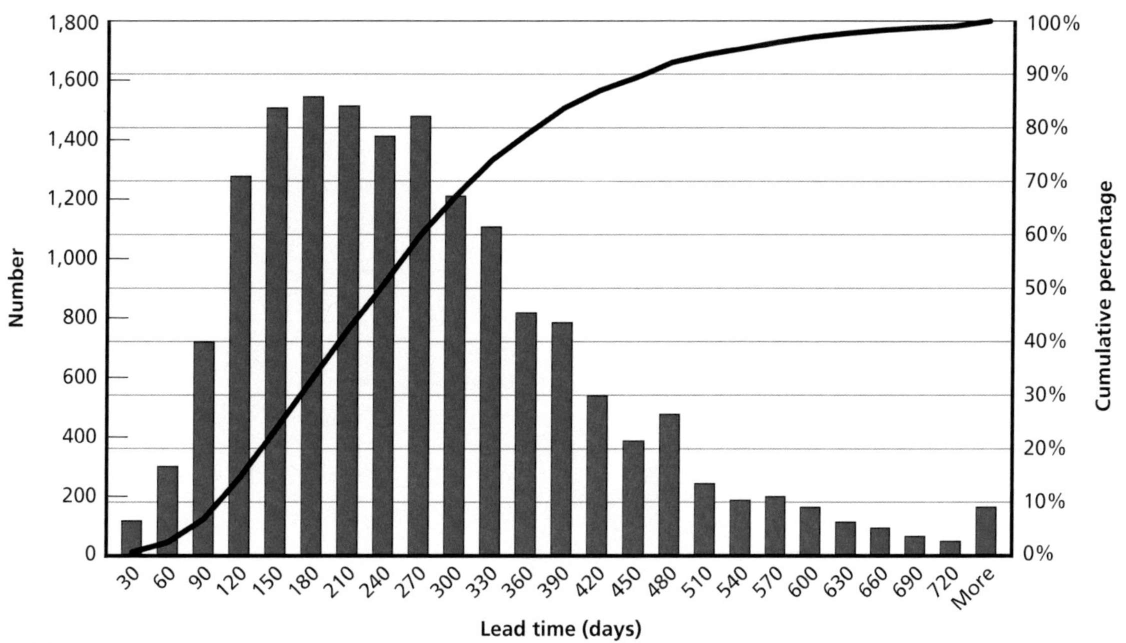

RAND *TR1274-6.1*

the items that drive DLA inventory value.[10] For these items, the median lead time, combining administrative or ordering lead time and production or supplier delivery lead time, is about 240 days, with close to 10 percent having lead times in excess of 720 days.

In conjunction with long lead times, DoD tends to order consumables in relatively infrequent large batches. DLA and the services use standard economic order quantity (EOQ) calculations to determine optimal order quantities but often apply minimum order quantities defined in terms of average days of supply, such as 90 days or 180 days. For more expensive items, this increases the order size above that which is purely economically efficient by increasing inventory on hand. This is done to limit procurement workload and avoid exceeding capacity limits for processing orders. The EOQ by definition is directly influenced by ordering cost, so it should already take workload costs appropriately into account, with high required workload per order driving up order quantities, too. However, the EOQ does not consider workload capacity, which may be the driving factor in using such minimums. DLA and the services also apply maximum order quantities, with two years being the maximum in DoD policy,[11] so as not to incur excessive risk in case the need for an item changes. This comes into play for very cheap items, which have high EOQs. Figure 6.2 shows the order quantities, called coverage durations, for the same set of items represented in Figure 6.1. The lower, blue, series indicates

Figure 6.2
Distribution of Coverage Durations for DLA High-Demand (in Dollar Terms) Items

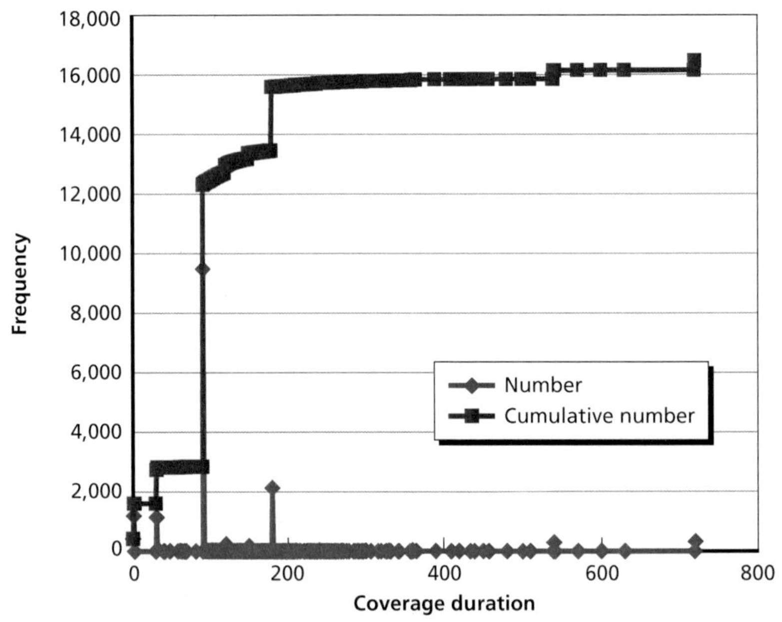

NOTE: The plot points were computed from a DLA enterprise business system query of coverage durations in December 2011.

RAND *TR1274-6.2*

[10] To create the sample, the top 20,000 National Item Identification Numbers (NIINs) (excluding direct vendor delivery items) in calendar year (CY) 2010 based upon extended value of demand were identified. Then these items were limited to those with four or more demands, for 16,469 items.

[11] DoD 4140.1-R.

the number of items with the indicated OQ on the x-axis. The upper, red, series indicates the cumulative number of items with the indicated OQ or less. One can see that rather than using a straight EOQ, most items have a default OQ, typically a minimum value, which would be above the EOQ.

As seen in the figure, 94 percent of the 16,469 items appear to have default values. The vast majority—80 percent of the total—of these are minimum values, primarily set at 30, 90, and 180 days of supply with some set to 31, 120, and 150 days of supply. Another 7 percent are set to a minimum of 1; these are mostly "non-forecastable" items with DLA replenishment method code (RMC) N. And another 2.5 percent have a value of 0. The coverage duration data were provided from a December 2011 snapshot, with the item set selected using 2010 demand data, so some of these may no longer be stocked items. This gets us to about 90 percent of the items being set to a minimum default value. The other 4 percent of set values are set at maximums of 540 or 720 days.

The Impact of Lead-Time Induced Forecast Errors

Through a series of graphs, this section shows the problems that long-lead times combined with high order quantities can create. Note that these examples are not average cases, but rather ones selected from items with inventory patterns exhibiting problems as opposed to the intended sawtooth type inventory pattern. In the graphs, which each have data for one item, the blue line shows monthly receipts of replenishments from suppliers into DCs (not when the orders were let) with the quantities indicated on the y-axis. The red line shows monthly demands, defined as the total quantity on the requisitions received by the supply organization during the month indicated (not when the stock was issued), and the green line shows an on-hand snapshot taken during the month. All of the examples cover January 2007 through September 2011.

In the first example, starting with Figure 6.3, demands are quite variable but relatively steady from January 2007 through September 2008. Replenishments are always just enough to keep some stock on the shelf, with inventory increasing in mid-2008 to a sufficient level to handle reasonable demand spikes such as the ones in February 2007 and August 2008. While developing a precise, accurate forecast for any given month is impossible, in October 2008, lacking any information from customers to the contrary, most forecast methods would project a little over 20,000 demands per month. This would have been an accurate, unbiased forecast. With a lead time of seven months, which would be at about the 40th percentile in the lead-time graph shown earlier, and an OQ of 90 days, in October the planners would be working on May through July 2009 and beyond.

As shown in Figure 6.4, in October 2008 demand dropped to just 3,500. But this was just one month and not that much lower than a couple of earlier low-demand months. Without intelligence on changes in customer needs, this would not be reason to change the forecast significantly. After a moderate uptick in November, demand then dropped below even the October level for several months. The question becomes at what point it becomes clear that the forecast should change. An overly responsive forecasting methodology would have swung wildly before October 2008, potentially suggesting the need for very large orders after demand spikes. This is problematic with long lead times, when these forecasts have to be used to plan far into the future and, with larger order quantities, for many months at a time. On the other hand, if demand does shift, one wants the system to respond as quickly as possible so as not to overbuy or underbuy. The longer the lead time, the worse the consequences of either over- or

Figure 6.3
First Example Inventory Pattern, January 2007–September 2008

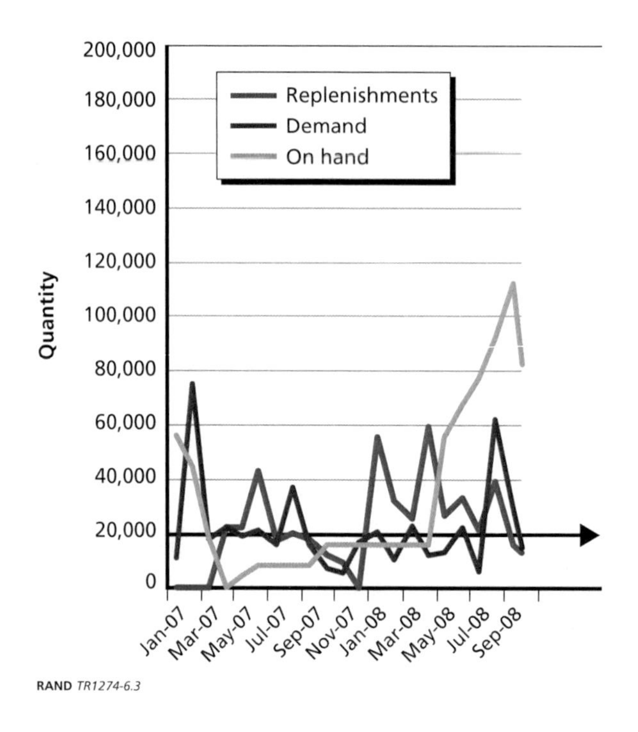

RAND *TR1274-6.3*

Figure 6.4
First Example Inventory Pattern, January 2007–September 2011

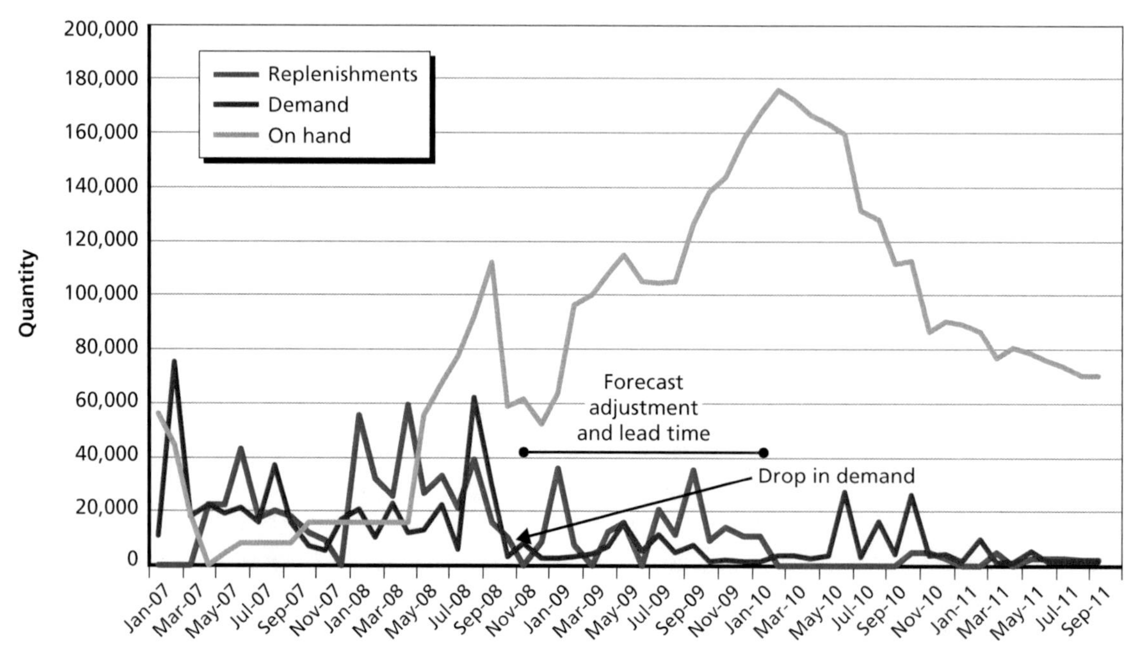

RAND *TR1274-6.4*

underreaction, so trying to find the right balance is critical but difficult. With a long lead time, there is high risk either way.

In this case, it looks like the forecast adjusted by the March to April time frame, with the last large order planned in April 2009. Deliveries often come in increments with the lead time representing the time to the first major delivery, so this led to substantial shipments through January 2010. With the benefit of hindsight, it is clear that the October 2008 drop represented the start of a new trend with a mean shift. But being confident of this much more quickly is not feasible through data alone. And we even see a short return to the former trend in June to October 2010, which would have added to the forecasting difficulty if it had occurred in early to mid-2009. The result, though, is that with the substantial forecast adjustment, order lead time, and OQ, orders remained much too high for much too long, building up substantial inventory. We do see that much of this was drawn down with the mid-2010 spike, but if demand continues at the late 2011 level, there will be several years of supply on the shelf.

The next example, shown in Figure 6.5, is similar in that a drop in demand combined with a long forecast adjustment and lead-time horizon, and a 90-day OQ leads to excessive inventory. What makes this case different is that there was a temporary demand increase that lasted long enough to "convince" the forecasting algorithm that it represented a real shift in demand, and given that it lasted 20 months, this would be a reasonable conclusion from the data alone. This further illustrates the difficulty of dealing with nonstationary demand patterns with high variability in the face of long-lead times.

High or excess inventory is one problem resulting from forecast error due to trend shifts on long-lead items. Another problem is stocking out, which can potentially impact readiness,

Figure 6.5
Second Example Inventory Pattern—The Impact of a Long but Temporary Demand Change

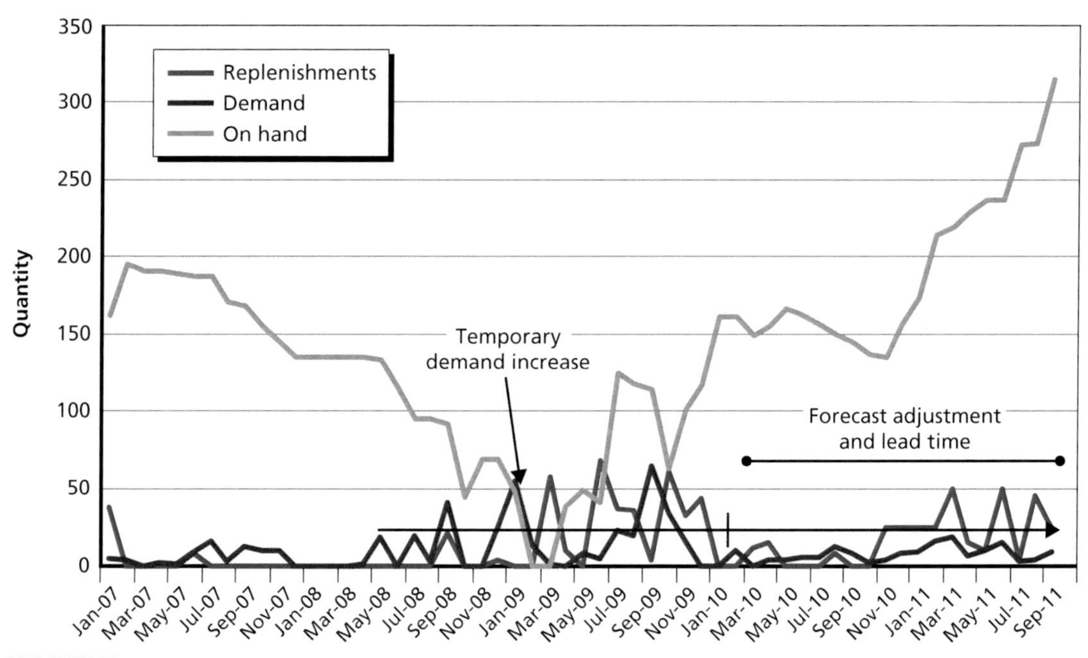

RAND *TR1274-6.5*

depending upon the item. And for this problem to occur, a full trend shift is not even necessary. Rather, one severe demand spike can create this type of problem. A single, large demand for a large quantity or a concentrated cluster of demands, which are not that unusual, can immediately drain on-hand inventory to 0. If such unexpected demands come in, the inventory planning system responds right away, not needing to wait to adjust the forecast in this situation. It will look at the new, much lower inventory position and determine the need for a new order considering any open replenishment orders and the forecast. However, even if the order is generated immediately, receipting it will not occur until a lead-time away. In the example in Figure 6.6, a demand spike in October 2007 led to a wiping out of on-hand inventory and a second demand spike occurred before the replenishment in response to the first one arrived.[12] Thus, the stock-out period ran from March 2008 until September 2009. In addition, the large procurements in response to these demand spikes ended up being excessive, leading to excess inventory.

Figure 6.7 illustrates how both problems can occur for the same NIIN in a short span of time, along with how severe these problematic inventory patterns and forecasting difficulties become in the face of very long lead times and highly variable demand. At first, there is seemingly excess supply. Then demand spikes drain inventory. After a brief recovery, another period of high demand again leads to stock-outs.

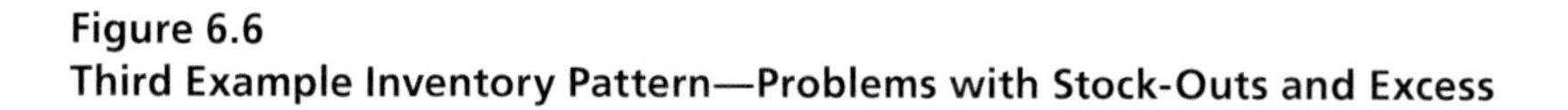

Figure 6.6
Third Example Inventory Pattern—Problems with Stock-Outs and Excess

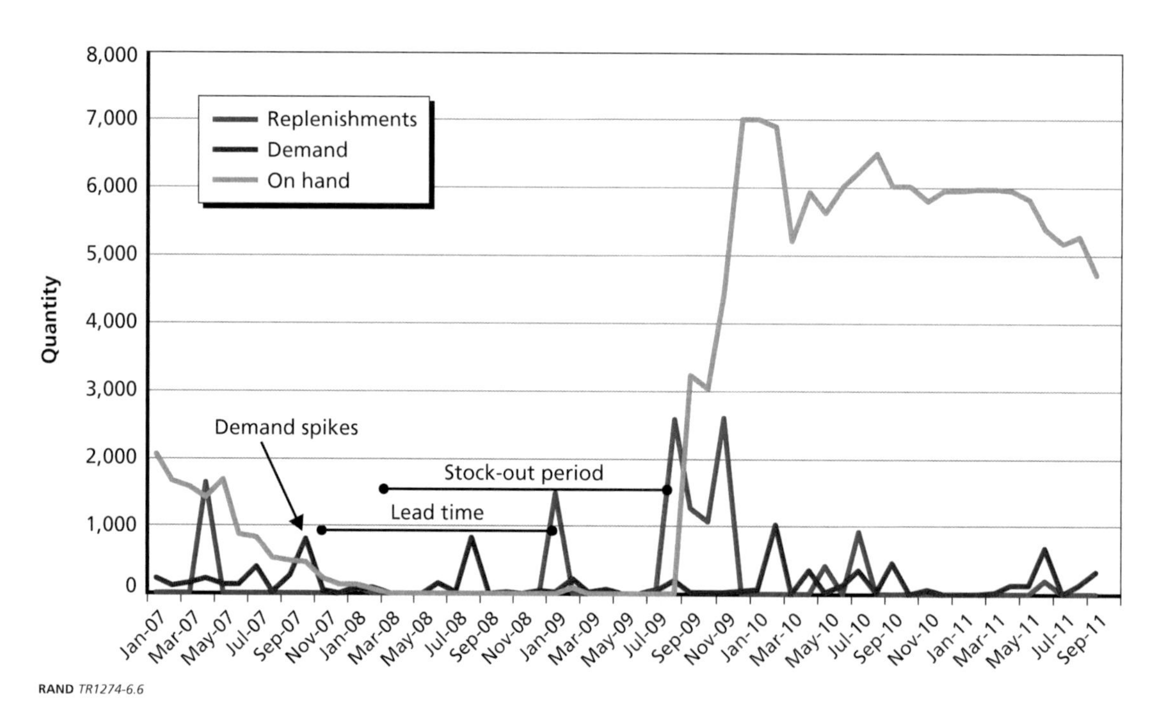

RAND *TR1274-6.6*

[12] The first demand spike was driven by two large orders in September and October 2007 from a maintenance depot. The planner filled a small portion of the orders immediately and then gradually filled it over time, with a substantial portion waiting until the January 2009 replenishment. This enabled smaller orders to be filled for a few months beyond this order. Similarly, planners sometimes begin rationing when inventory gets low to continue to fill orders for high-priority customers and/or requisitions.

Figure 6.7
Fourth Example Inventory Pattern—The Impact of Very Long Lead Times

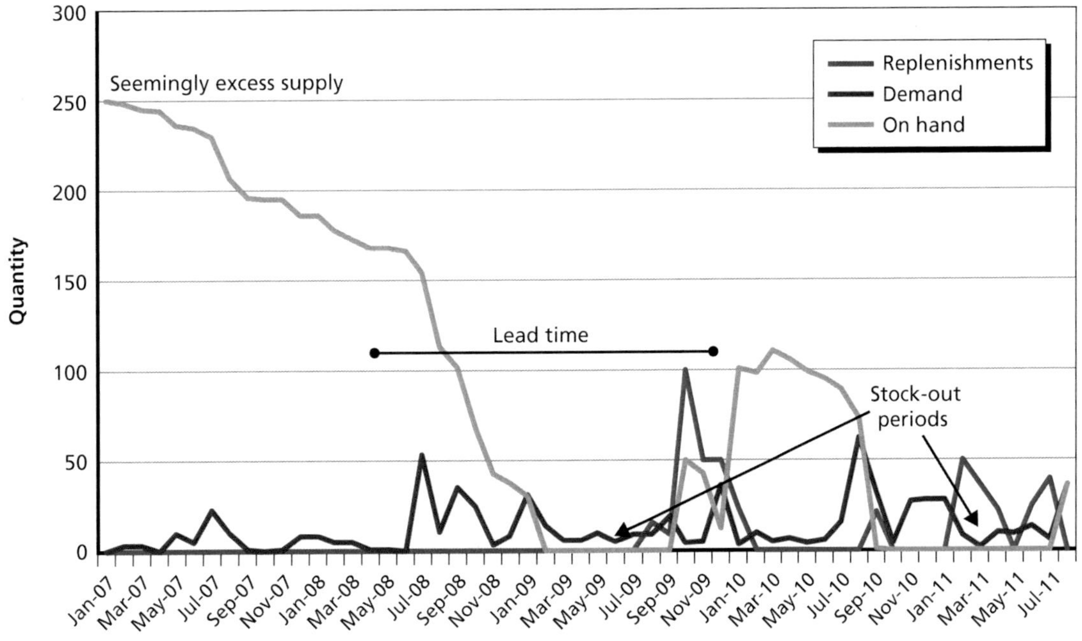

RAND *TR1274-6.7*

The next example, in Figure 6.8, shifts to a focus on order quantities. For this item, two large orders were placed in late February and another in May of 2008. In relative terms, the planning lead time for this item is relatively short, at about four months. The four-month planning lead time, though, is for the first significant delivery, as is the case with all planning values. These orders were typically filled in quantities of 10,000 per month, so these orders extended through a planning horizon that reached November 2009. But in January 2009, demand abruptly dropped and stayed very low. Inventory for this one item piled up until the orders were completely filled by the supplier, with inventory eventually reaching about $22 million. In some cases, contracts can be cancelled, but it depends upon the terms and conditions, and, even if feasible, it may not be economically beneficial, depending upon any penalties in the contract.

In contrast, the final example shows the inventory benefit of cutting order quantities. In Figure 6.9, the second and third replenishments are less than half of the first large replenishment in this time period. As a result, the average on hand goes down by about one-third, and there is less inventory at risk if there is a decline in demand. Note that this is also an illustration of an item with stable but highly variable demand. The stability enables a classic sawtooth inventory pattern without excess supply or stock-outs, in sharp contrast to the other examples that illustrate these problems.

To support an analysis to understand the effects of lead times and other factors on inventory levels and performance, DLA provided requested data for a large sample of items that drive the dollar value of its inventory requirements and purchases. The sample consists of the top dollar drivers in terms of demand in CY 2010 with at least four demands, which resulted in 16,469 items. We conducted a statistical analysis of this sample in a way that would allow for determining the relative impact of factors that affect inventory levels and customer service

Figure 6.8
Example Impact of Very Large Order Quantities

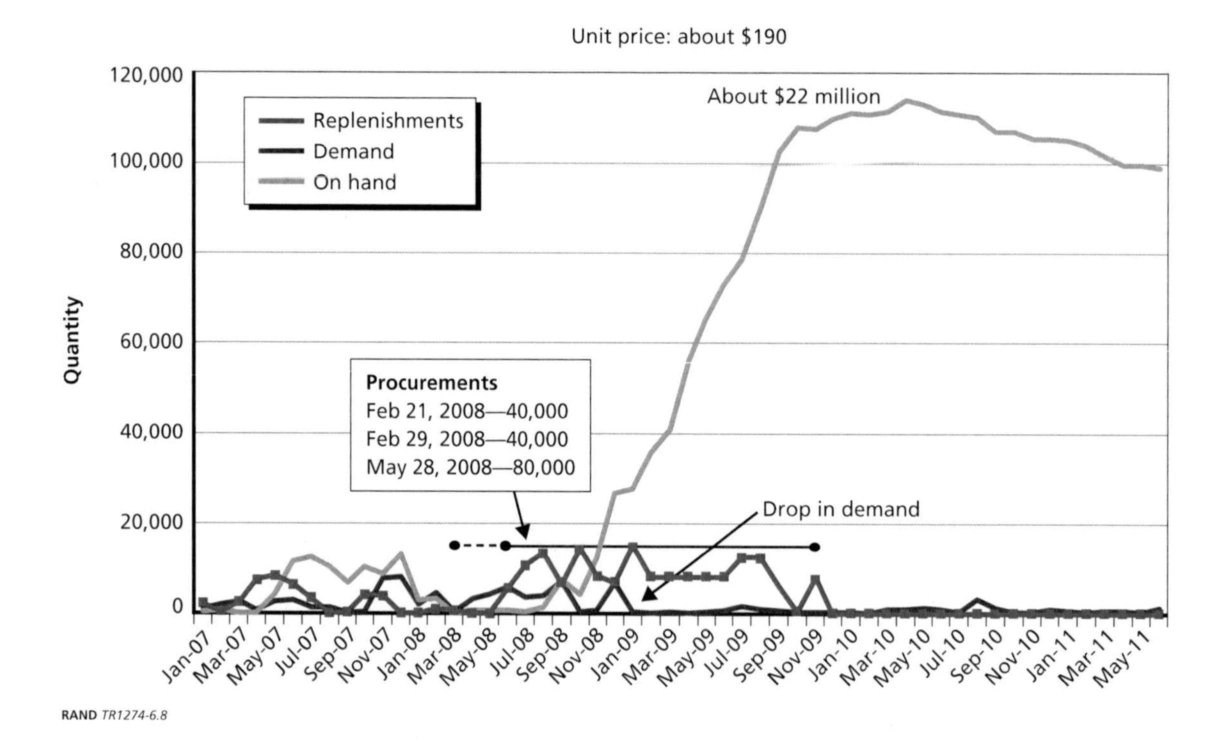

RAND *TR1274-6.8*

Figure 6.9
Example of the Impact of Reducing Order Quantities

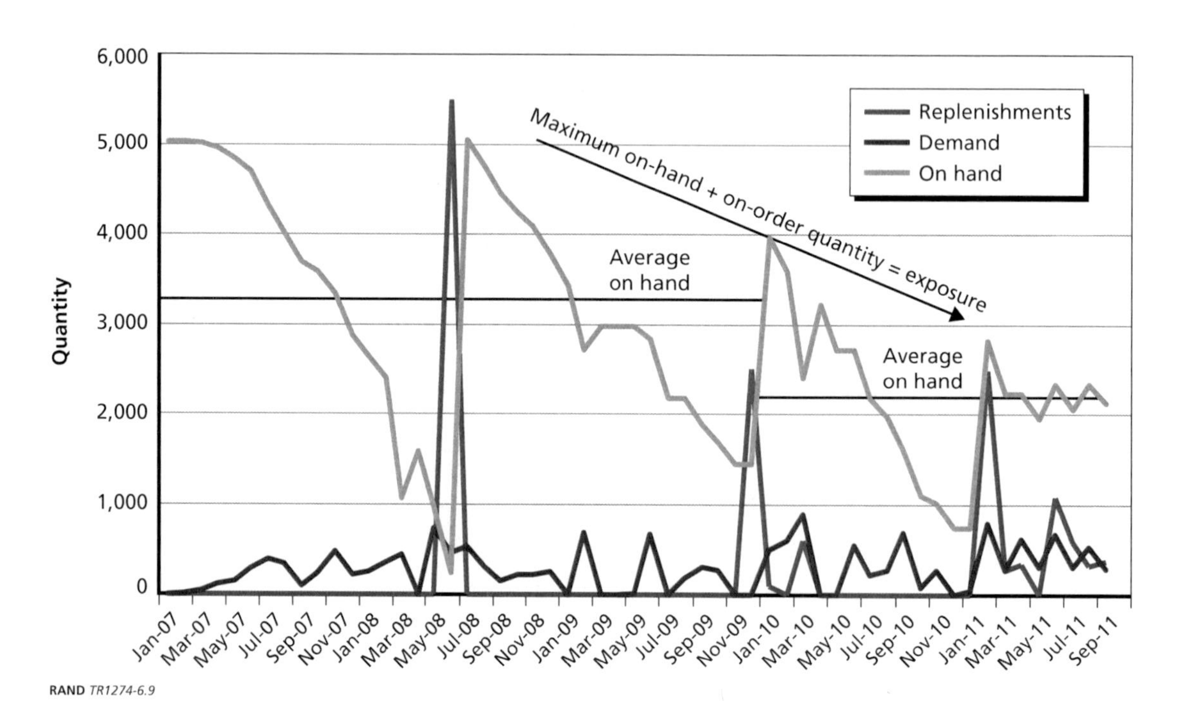

RAND *TR1274-6.9*

and that would enable the determination of the effect of each factor when controlling for all of the others.

The factors examined consist of item characteristics and inventory management parameters. The item characteristics or givens include unit price, demand variability, demand level, whether or not the item has a single dominant customer, the DLA supply chain management organization that manages the item, and the lead time,[13] with the latter being somewhat controllable depending upon supplier selection, negotiation, and supplier management and performance. The management parameters are primarily yes or no indicators, representing whether or not the associated method is applied. These include whether or not the item is a collaborative planning (collaborative) item, a special program requirements (SPR) item, a forecastable item (based upon RMC), or an initial provisioning item. The last factor is the forecasting model used for the item. Collaborative items are ones in which customers provide automated data feeds of their forecasted needs based upon usage plans; SPR items are those for which projected usage based upon production plans is provided; the RMC designation determines whether a forecast is used for supply planning or whether an inventory minimum and maximum are used; and initial provisioning items are stocked in accordance with engineering forecasts. The order quantities or coverage durations were not used since they are set to default values for most items.

To gauge performance, we used two traditional measures and two new ones to find the types of patterns found in the examples and that are indicative of situations when there might be excess inventory to be disposed of or readiness problems. The traditional measures are the backorder percentage (often called the backorder rate), which is 100 percent minus the materiel or stock availability percentage, and inventory turns. The two new ones are the longest stock-out period in weeks and the maximum months of supply on hand.

Among the potentially controllable or management factors, the only one with a high effect is lead time. The other three are unit price, demand variability, and demand level. Unit price has the strongest effect, which is consistent with the logic used to set safety levels. While working within some constraints, such as a minimum service level for items, DLA aims to meet material availability goals at minimum inventory cost, which translates into the allocation of safety levels and associated service-level targets. The more expensive an item is, the lower the service level and safety stock will be. This should result in a higher backorder rate, higher inventory turns, more extended stock-out periods, and smaller inventory peaks the higher the price of an item is, which is what the statistical analysis found. The safety level should primarily be a function of the relative demand variability of an item and its unit price. What is interesting is that even controlling for unit price, which affects the safety level, demand variability is still a very significant factor. The lead-time effect is on par with the effects of demand variability and demand level.

The four graphs in Figure 6.10 show the estimated effects of lead times with respect to the four metrics. In the graphs, one should key on the general patterns and range of effects rather than the precise lines. Additionally, confidence in the estimates at the tails of the graphs is lower with the records more spread out in terms of lead times (the small tick marks represent percentiles in multiples of 10 for the lead times in the sample). The upper left graph suggests that very low lead-time items have low backorder rates, with a continual increase in backorder

[13] These include land, maritime, aviation, clothing and textiles, construction and equipment, medical, and subsistence.

Figure 6.10
Estimated Effects of Lead Times on Inventory Performance

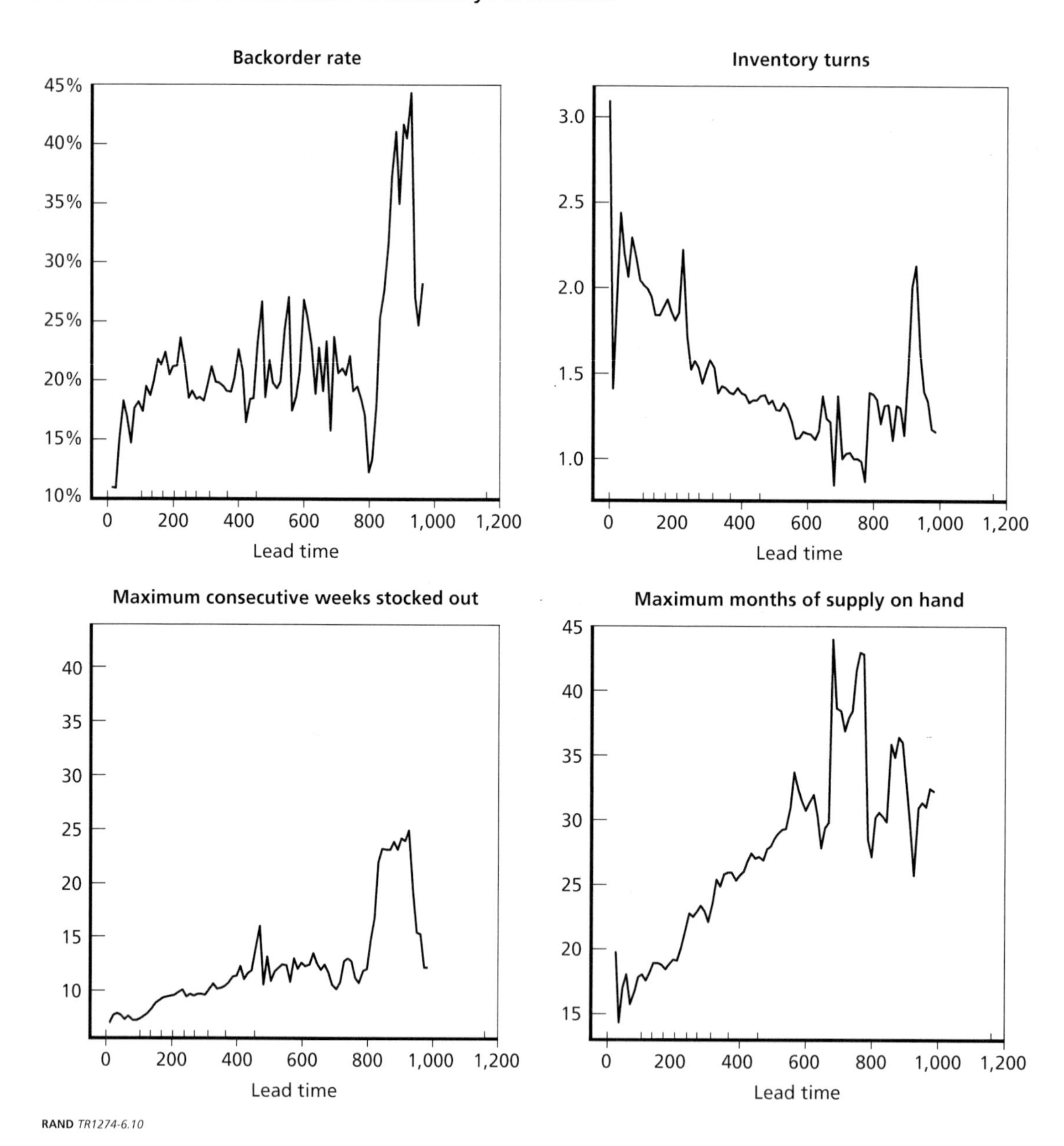

rates up to about 200-day lead times. Then very high lead-time items tend to have much higher backorder rates. The overall range is from the low-teens to about 45 percent—a substantial range and effect. The pattern seen for the maximum consecutive stock-out weeks (lower left) is similar, as one might expect, with a range from a little less than 10 weeks for low lead-time items to 25 weeks for long lead-time items. Correspondingly, inventory turns (upper right) are higher at about 2.5 turns per year for short lead-time items than for long lead-time items, with turns of about 1. The maximum months of supply (lower right) has the opposite pattern from turns, as it should, with a range from about 15 to 40 months.

The supply chain management organization of the item, which could also be a proxy for the type of item, was found to have a moderate level of effect, indicating that controlling for the other factors, practices in planning and inventory management differ across the DLA supply chains. In particular, the subsistence, and to a degree clothing and textiles, supply chain stands out from the other on these four performance dimensions after controlling for the other factors. The forecast model used for an item is the last factor with a meaningful effect, although at a lower level effect than for the supply chain and with much less effect than for the high-impact factors discussed earlier. The other factors all were found to have relatively little to no effect in the statistical analysis.

This suggests that either collaboration is not done well, the information that the customers provide is no better than historical data, and/or that the customers face equally difficult forecasting and variability problems in terms of their usage and production plans. For example, depot end item production plans could change frequently or be planned far in advance and then subject to some change in execution. Such changes in depot production plans would lead to similarly unanticipated shifts in the demand for the associated repair parts.

Appendix B shows the relative effects, on a 100-point scale, and the range of effects for each factor, along with an explanation of the statistical method employed.

Consistent with the discussion of inventory on hand versus what might be expected given safety levels and order quantities, this statistical analysis and the examples suggest that lead times are a much bigger contributor to on-hand levels than sometimes considered to be. The examples and the statistical analysis show how forecast error from trend shifts becomes greater and leads to more excess as lead time becomes greater.

Despite this criticality of lead times and order quantities, neither has received significant, sustained attention at the merited level by GAO, the Office of the Secretary of Defense (OSD), the services, and DLA. This lack of attention can be seen in reports that talk about inventory management problems, which emphasize different solution paths to reduce inventory; the DoD Comprehensive Inventory Management Improvement Plan (CIMIP), which had to address section 328 of the National Defense Authorization Act (NDAA) for FY 2010 and which in turn reflected prior GAO concerns; the lack of metrics and associated management and command attention on lead times and order quantities; and a lack of demonstrated improvement.

The following examples illustrate this divergence in emphasis from lead times and order quantities. In 2011, GAO stated "Our recent work identified demand forecasting as the leading reason why the services and DLA accumulate excess inventory."[14] Albeit to a lesser degree, this same report did also recognize the need to address lead times, which GAO has previously emphasized. In the CIMIP, there are eight primary actions that are the focus of the plan, each addressing one of the eight required elements of the 2010 NDAA, and a set of other actions that were added and listed as the ninth category. Improving lead times is one of the four "other" actions.[15] In supply chain metrics provided by the services and DLA in mid-2011, only one of the five were tracking lead times, although two stated that doing so was in development. And while lead times have not been tracked at the OSD level, the metrics working group of the

[14] GAO, Defense Inventory, *Opportunities Exist to Improve the Management of DOD's Acquisition Lead Times for Spare Parts*, Washington D.C., GAO-07-281, March 2007.

[15] Assistant Secretary of Defense for Logistics and Materiel Readiness, *Comprehensive Inventory Management Improvement Plan*, October 2010.

Deputy of the Assistant Secretary of Defense for Supply Chain Integration intends to include lead times in new OSD supply chain enterprise metrics.[16]

This is not to say that improving these processes has not received attention or that actions to improve them have not been taken. Rather, significant efforts have been made across DoD to implement improved purchasing and supply management practices by adopting new commercial practices in strategic sourcing and supplier relationship management. Among other effects, these efforts should lead to shorter lead times and lower order quantities. Yet, we were unable to find any evidence of shorter lead times and lower order quantities as a result of service and/or agency strategic sourcing and supplier relationship management efforts. In some cases, the metrics have not been tracked, in others the historical data availability is limited to a short period of time, and in others the data are considered suspect.

In the course of this project, we were able to obtain metrics, or data to compute metrics, to examine relatively recent trends for two services/agencies. Both actually show small increases in lead times since 2008. Figure 6.11 shows the per order averages (unweighted) from 2008 through 2011 (through October) for one DoD organization, with the left graph showing administrative lead time (ALT), production lead time (PLT), and overall lead time. The graph on the right shows the averages weighting each transaction by its dollar value, more directly tying the metrics to the effects on the inventory.

The trends are consistent with findings that DoD has reduced its emphasis on lead-time reduction. In 2007, GAO found that from 1994 to 2002, DoD had been relatively success-

Figure 6.11
Lead Times for One DoD Organization

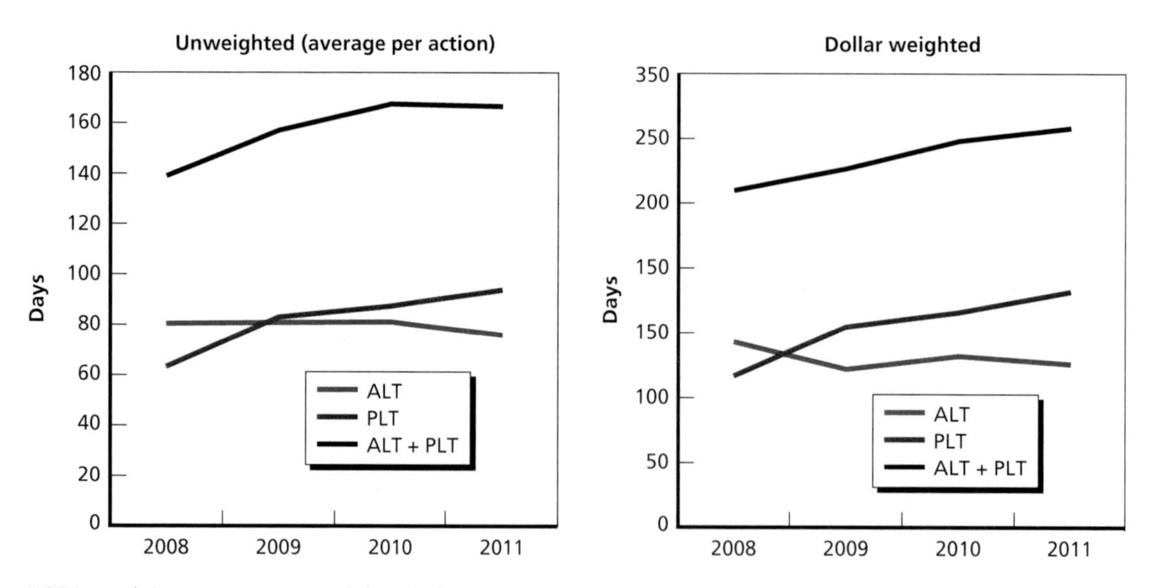

NOTE: Lead times were computed directly from the organization's transactional data.
RAND *TR1274-6.11*

---

[16] Office of the Deputy Assistant Secretary of Defense for Supply Chain Integration Supply Chain Metrics Group, "Proposed Enterprise Metrics as of 20 Dec 11," December 20, 2011. U.S. Naval Supply Systems Command, "NAVSUP Business Metrics Review Session Logistics Support," December 22, 2010. Department of the Army, "Army Metrics Submission," June 2011. U.S. Air Force Global Logistics Support Center, 2011. U.S. Marine Corps Logistics Command, "CSCMP Marine Corp," May 2011. Defense Logistics Agency, "Agency Performance Review," February 2011 and September 2011.

ful at reducing lead times, achieving an average reduction of 5.6 percent per year through "streamlining internal administrative processes, oversight from USD(AT&L) [Under Secretary of Defense for Acquisition, Technology, and Logistics], and developing strategic relationships with suppliers."[17] However, from 2002 to 2005, GAO concluded that significantly reduced OSD emphasis on lead-time reduction, such as the lack of goals, metrics, and reporting on improvement efforts, led to a falloff in improvement to 0.9 percent per year. Despite, the reduced OSD oversight, GAO did note new DLA and Air Force initiatives over this period, with some resulting reduction in lead times compared to no improvement for the Army and the Navy.[18]

It is quite plausible that focus shifted to planning for OIF, then to resolving the myriad distribution and supply problems that began with the start of major combat operations, and then to responding to continually shifting requirements as operations continually evolved. While distribution problems were largely resolved by the end of 2003, it took longer to address supply problems and then planning efforts were needed for a variety of new requirements from reset of equipment to ever changing events on the ground such as the surge in 2007. Interactions with a variety of representatives from across the services suggest that the prior focus had yet to return as of 2011, and neither the DLA nor the Air Force data show improvement in lead times since 2008.

Notably, demand forecasting, supply performance, and inventory management and performance are typically considered the purview of item managers or demand and supply planners, who focus on setting forecast methods, determining levels, and determining when orders need to be placed. They, to a large degree, use lead times as inputs, along with order quantities as the result of supplier constraints or procurement workload limits. Yet, it is these latter parameters, negotiated and managed by procurement and supplier management personnel, that drive inventory levels. These personnel tend to be more focused on item prices and suppliers meeting commitments than in selecting suppliers based on lead times and then focusing on ways to improve lead times and order quantities. In short, they have a large effect on inventory levels and performance without equal responsibility for and attention to inventory levels and performance. If one applies the guiding supply chain design and enabling mechanism principles discussed in Chapter Five, it is not clear that their decisions and efforts sufficiently reflect total cost and supply chain performance effects. Nor do procurement organizational metrics and accountabilities fully reflect the full set of outcomes and costs that these personnel significantly influence.

Thus, this analysis leads to the recommendation for a new, holistic DoD initiative to examine how best to reduce lead times and order quantities while considering total costs to ensure they are not decreased at the expense of other cost categories. This should potentially encompass (1) how DoD selects, manages, and collaborates with its suppliers; (2) demand and supply planning practices; and (3) organizational design, capabilities, metrics, and accountabilities. With regard to suppliers, a likely important area to examine would be whether infor-

---

[17] We note that lead times we measured for using the raw actual data are substantially higher than what GAO shows for 2002 and 2005 and even 1994 for the same service or agency. Notably, GAO relied on budget stratification data, which use planning lead times. Although, GAO did compare actual delivery times to estimated times for this organization, and actually found that it had overestimated lead times (GAO, 2007).

[18] GAO, 2007.

mation flow improvements from DoD to its suppliers could improve their planning, affecting lead times and order quantities.[19]

Reparable Inventory

While working with suppliers to reduce lead times and order quantities should have a significant effect on consumable item inventory, largely managed within DoD by DLA, much of the dollar value of DoD's secondary item inventory is in reparable items, so an inventory discussion would be remiss in not including this portion of secondary item inventory. As shown in Figure 6.12, in FY 2011 almost $64 billion of the $85 billion in secondary item inventory stored in DLA DCs was for reparable items, with over half the reparable inventory being unserviceable. In terms of the dollar value, each reparable item turned about once every 2.6 years.[20] On average, this means that from the time a broken reparable is receipted back into the wholesale supply system, it takes 2.6 years for induction, repair, and issue as serviceable.

In the short term, improving processes cannot significantly reduce this inventory unless some of it is disposed of. Normally, it is reduced only at the rate at which broken items are condemned as not economically reparable. What can be done, though, is to reduce initial and subsequent buys of new reparables so that there are fewer in the system.

Initial buys should be based upon the projected closed loop retrograde and repair times, which drives the amount of inventory that needs to be in the system. At times though, if this

Figure 6.12
Inventory Turns by Item Category and Condition

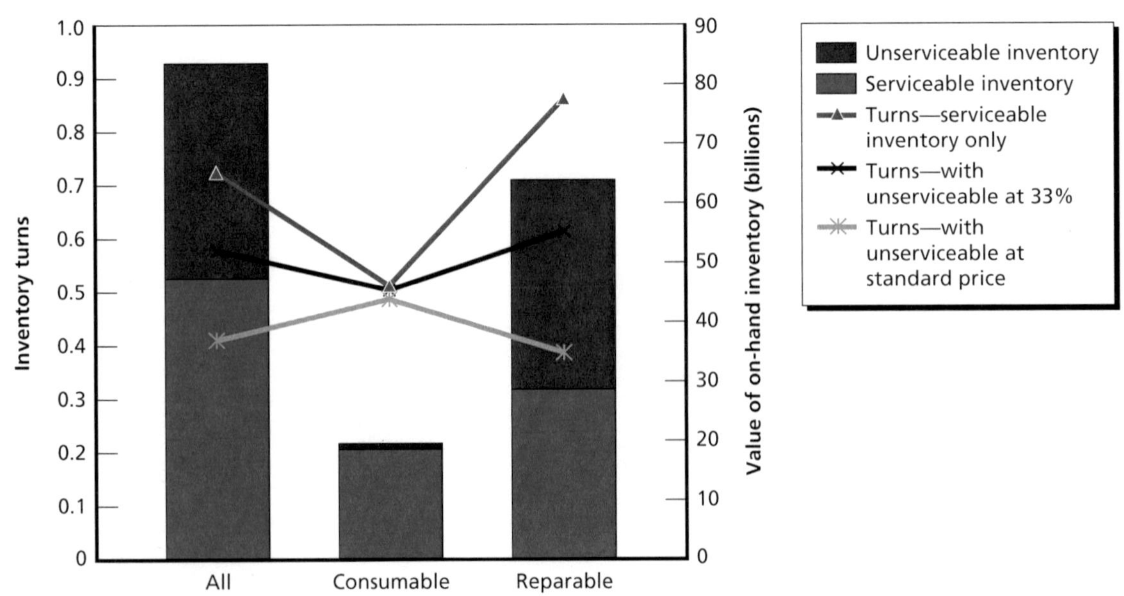

SOURCE: The figure data are based on data from the Strategic Distribution Database and the QBO.
RAND *TR1274-6.12*

[19] ASD(L&MR) initiated such a study in April 2012 focused largely on DLA, which manages most DoD consumable items.

[20] Based upon Strategic Distribution Database data.

process gets disrupted, more may have to be bought to meet demand. Or if demand shifts without a corresponding shift in the repair plan, then more may have to be bought to meet demand.

Thus, there is the need for supply chain integration in the planning of reparable item inventory and repair on three dimensions. The first dimension is linking reverse distribution to repair and reparable item inventory planning. Improvements in this process can lead to lower requirements for new reparables, and they can enable reduced inventory of existing reparables, with "washouts" not having to be replaced until the total system inventory hits a new, lower equilibrium point. The second dimension is ensuring that no disruptions in repair or the retrograde of carcasses occurs or that they are minimized in severity. Such disruptions are possible from problems in the reverse pipeline that returns carcasses, repair budget shortages, supporting parts shortages, and repair capacity shortages. The first of these potential disruptions happened early in OIF, leading to the need to buy more of some reparables when repair was first starved of sufficient carcasses to repair and then could not produce at the necessary "makeup" rate to dig out of the serviceable inventory hole. Avoiding the second and third potential disruptions requires integrated considerations of repair and supply budgets and planning. Avoiding the fourth requires integrated consideration of repair capacity, end item maintenance planning, and sustainment demand planning.

The third dimension is ensuring that repair plans and demand are kept in sync. Shifts in demand with delayed shifts in repair plans can lead to serviceable item shortages that have to be made up through purchases. Conversely, such misalignments can also lead to excess production, resulting in excess serviceables. This is less a supply problem, than one of excess repair capacity used, increasing maintenance costs or creating an opportunity cost if another item's needed production was forgone.

With detailed Army repair production and demand data, we have seen 9 to 12 months' lags in trend shifts—both higher and lower—in demand prior to shifts in production. This stems from an annual planning process that locks in repair schedules in the third quarter of the prior FY. There is some adjustment, but it is limited during the year of execution. Shifting to more of a pull production system would eliminate much of this lag. In earlier research, RAND researchers posited that quarterly updates would eliminate much of the over- or underproduction without overreacting to the underlying variability in demand.[21]

Examination of Air Force depot production patterns and processes reveals that the same problem does not exist in the Air Force. Production is planned based upon demand in the spirit of a pull system, with business logic that prioritizes the use of repair capacity and funding to meet the highest-priority needs. We do not see the same lags in responding to trend shifts. For some items, no or very little Air Force serviceable inventory is kept on the shelf, but as soon as a demand comes in, the item is produced relatively quickly. However, for some items, there are also orders of magnitudes more of some reparables in the system than needed. It is hypothesized that this may have been due to fleet size reductions, but this has not been investigated. Another possibility is that processes have improved since the original purchases were made. Regardless, given the current production planning process in the Air Force, the primary opportunities for reducing future supplies of Air Force reparables would be in further

[21] Wang, Mark, Jason Eng, Rachel Rue, and Jeffrey Tew, "Adapting Secondary Item Planning to Pull Production," unpublished RAND Corporation research, 2009; and John R. Folkeson and Marygail K. Brauner, *Improving the Army's Management of Reparable Spare Parts*, Santa Monica, Calif.: RAND Corporation, MG-205-A, 2005.

integrating retrograde processes and spare parts planning for depot repair to reduce the total cycle time from return to repair to minimize future buys and understanding what drove long positions of some items to put in place practices to avoid this from happening as new items are introduced into the system.

Stemming from these practices and observations, Army reparables have less backorder time on average than Air Force reparables, with Air Force reparables having higher inventory turns. Navy and Marine Corps reparable backorder times are between the two, with turns similar to those for Air Force reparables for the Navy and similar to the Army for the Marine Corps. Additionally, an initial review of the data for Navy reparables suggests a system closer to the Air Force pull-like model than the Army's. Still these different backorder time and inventory turn relationships suggest some differences in processes for the four services, although we have not examined Navy and Marine Corps reparable management processes.

Examining how the service materiel and/or system commands could improve how their demand and supply planning organizations work with depot maintenance and DLA, financial planners, and operational planners to reduce the need for reparable item inventory and new buys is essential to having a dramatic impact on DoD inventory. Tightening the integration between demand and repair planning, using more Air Force pull-like processes with some adaptations or adopting the type of pull-like production system suggested for the Army, could offer some opportunities to prevent future buildup of excess reparables. Further tightening the linkage between retrograde processes and reparable inventory management and production planning is another avenue to explore. Another would be financial control processes, with some reports that difficulty in shifting working capital fund obligation authority during the course of an FY contributing to some of the friction in repair plan adjustments.[22]

---

[22] ASD(L&MR) initiated a study in April 2012 to examine depot-level reparable item management encompassing an examination across all four services.

CHAPTER SEVEN

Scheduled Trucks—Apply a Systems View for Shipment Consolidation

In the mid-1990s, in collaboration with the Army as part of its Velocity Management initiative, DLA instituted scheduled trucks from its SDPs to major Army installations. These allowed full-truck-load-like rates with express-delivery-like service. Prior to scheduled trucks, shipments from an SDP to an installation were shipped by different modes, depending on the priority. By consolidating shipments for an installation across all priorities on a periodic basis, a lower shipping rate could be achieved for all shipments, even low-priority shipments.[1] Additionally, within an installation, a truck can stop at supply activities in a standard order at scheduled times, enabling improved receipting times and productivity, which is also aided by one delivery per day rather than multiple deliveries from different transportation modes. The greater the volume that can be consolidated, the lower the average shipping cost per pound and the more frequent the trucks, leading to faster delivery times. Thus, scheduled trucks work best when the facing fill—the percentage of shipments from the designated first source of materiel for a customer—from the supporting SDP for a truck's customer set is high and as much of the material sent to an installation as possible is sent via the truck. Through the Strategic Distribution Management Initiative and continuing efforts by DLA Distribution, the scheduled truck network was expanded to cover most large installations in CONUS.

To demonstrate this benefit, Figure 7.1 shows the requisition wait time (RWT) in days and shipping cost per pound for different shipping modes from DLA DCs to customers in CONUS in FY 2011. Each column shows the RWT, with the lower black portion of the columns showing the median time, the middle yellow portion of the columns showing the 75th percentile time, the upper grey portion of the columns showing the 95th percentile time, and the red squares indicating the mean times. The blue triangles indicate the cost per pound using the right y-axis scale. Unless there is a single very large order, which can fill a full truck, shipments are sent via five primary modes. Low-priority shipments are shipped via small package surface carriers or via LTL, depending upon the size. Both have much longer RWTs than scheduled trucks with LTL being a little more expensive and small package shipments much more so. High-priority shipments are shipped via overnight express air service, such as FedEx or UPS, or commercial air freight, depending upon the size and weight. Overnight air times are similar to scheduled truck times but are close to an order of magnitude greater in cost, with commercial air freight being a little slower and also much more expensive. These comparisons are to the average scheduled truck cost and performance, although execution quality and the

[1] Mark Y. D. Wang, *Accelerated Logistics: Streamlining the Army's Supply Chain*, Santa Monica, Calif.: RAND Corporation, MR-1140-A, 2000.

Figure 7.1
Cost and Performance of Different Transportation Modes in CONUS

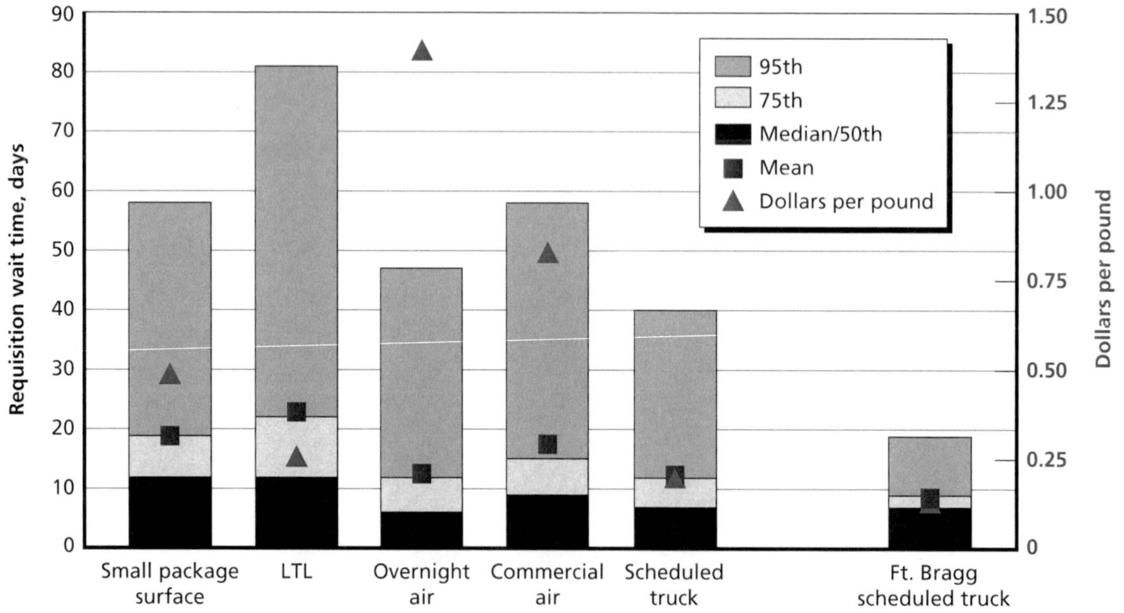

NOTE: The source for this figure and all of the others in this chapter is the Strategic Distribution Database.
RAND *TR1274-7.1*

associated performance and cost metrics vary for service to different installations. With high scheduled truck shipment volume from strong participation across the installation and a one-day drive from its supporting SDP, service to Ft. Bragg has even lower costs and significantly less variability in times, as shown in Figure 7.1, which includes Ft. Bragg as a benchmark.

However, the current scheduled truck network is suboptimized. As this discussion will show, there are opportunities for improvement by applying several of the guiding principles: leveraging scale advantages, ensuring metrics reflect downstream influences of decisions, aligning decision rights with the system view, ensuring organizations are accountable for downstream costs they influence, and having necessary decision support tools.

These factors lead to two primary shortfalls: (1) There are some supply activities on scheduled truck installations that are not served by the scheduled truck, and there are other supply activities that are served by scheduled trucks for lower-priority shipments but have high-priority shipments sent via the two air modes; (2) there are some moderate-size installations that do not have scheduled truck service. Figure 7.2 shows the percentage of shipment weight sent via scheduled truck to the top 25 destinations, by weight, for shipments from Defense Distribution Depot Susquehanna, PA (DDSP), in CY 2011.[2] The columns show the weight shipped (left y-axis), and the line series shows the percentage of weight on scheduled trucks. The installations with the highest percentages are at about 90 percent. Those in the 50 percent

[2] DDSP is the SDP in the eastern half of CONUS that serves operational customers. It previously supported all FDPs in the east, although in 2011 the designated primary replenishment source for FDPs in the southeast was shifted to Defense Distribution Depot Warner Robins, GA, designated an SDP as part of BRAC 2005, and Defense Distribution Depot Oklahoma City, OK, at Tinker Air Force Base was redesignated as an SDP as well. However, as of 2011, all installations in Figure 7.2 continued to receive significant amounts of materiel from DDSP.

Figure 7.2
Percentage of Shipments from DDSP to Its Highest-Volume Customer Installations on Scheduled Trucks, CY 2011

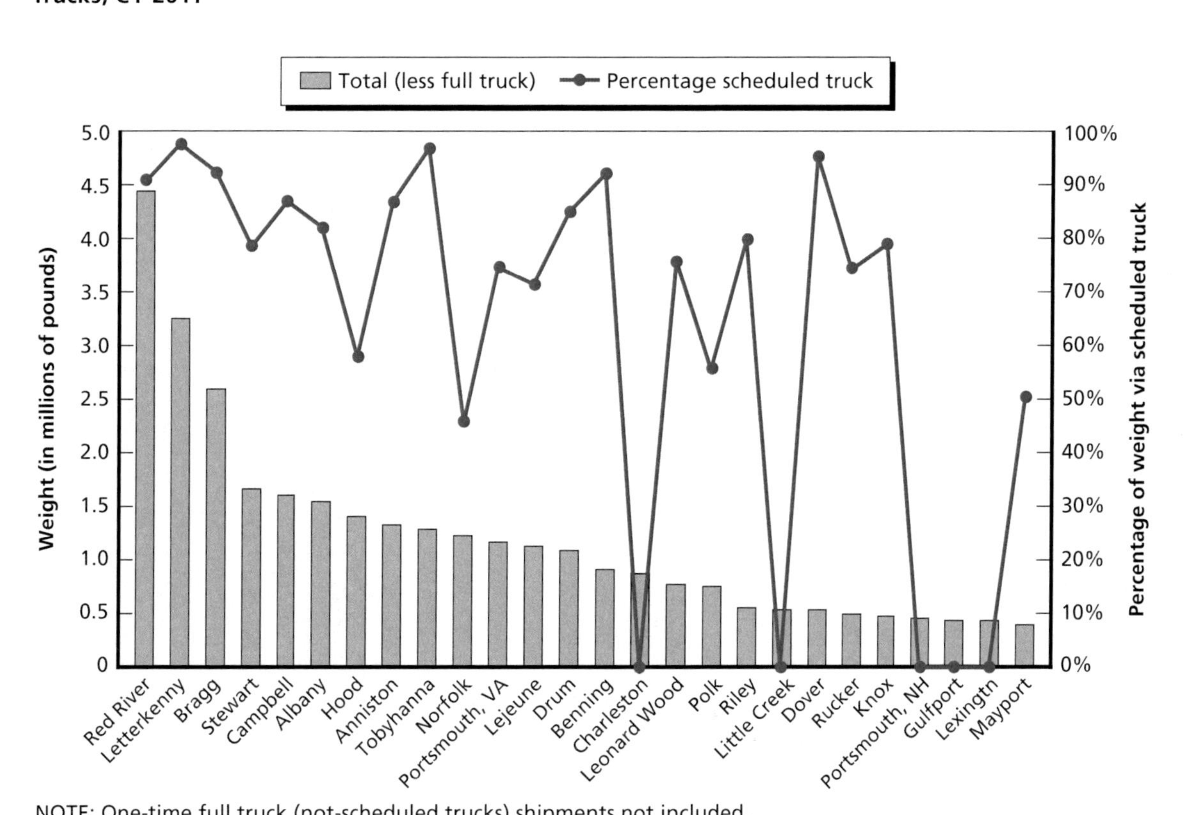

NOTE: One-time full truck (not-scheduled trucks) shipments not included.
RAND *TR1274-7.2*

to 60 percent range have substantial "leakage" from the scheduled truck from DDSP to their installations—or in other words, they have multiple supply activities not on scheduled trucks or that require high-priority shipments to be sent via air. Those with 0 percent indicate installations without scheduled truck service.

The leakage arises from a couple of reasons. The first reason is that it is ultimately up to each supply activity how it wants to be served, with potential influence from the installation or command to which it belongs. A supply activity can ask to not be served by scheduled truck or can ask that the truck be used for only some priorities. The second reason is that a supply activity and its supporting unit may be new to the installation and DLA Distribution has not coordinated with it to add it to the scheduled truck service since it was restationed. Service can be initiated in two ways: The supply activity can contact DLA Distribution or DLA Distribution can contact the supply activity. However, there is not a standard, systematic process within DLA Distribution for determining which customers should be added to routes or which routes should be added that automatically result in coordinating with new, unserved customers on an installation.[3]

[3] The Army is in the process of assigning the coordinating role on the Army's end to Army Materiel Command Installation Supply Representatives (ISRs). The ISRs will be intended to identify non-scheduled truck supply activities that should receive scheduled trucks and to then coordinate appropriate changes with DLA. Additionally, they can serve to coordinate scheduled truck plans with DLA Distribution.

When such leakage is identified, though, it can be addressed quickly, generally improving service and lowering the cost to serve the supply activities newly added to a truck route as well as potentially for the other customers on the installation. The overall cost per pound will come down as more weight is sent on the existing trucks. If the new customers increase weight sufficiently to increase truck frequency, than service for all customers will improve. Figure 7.3 shows the benefit that occurred for two supply activities that previously had not been served by the Ft. Bragg truck when they were added to the truck service. The left column shows the RWT and cost per pound for the two prior to being served by scheduled truck, and the right column shows the same metrics when served by the scheduled truck. RWT improved by 3 days on average, with a 15-day improvement at the 95th percentile, and the average cost to serve them dropped from $0.34 per pound to $0.12 per pound.

The lack of a standard route planning process also contributes to the problem of installations not being served by scheduled trucks at all. Many scheduled truck routes service one installation. There is not a standard process to identify multi-installation routes; nor is there an automated planning tool to identify the best potential set of such routes. Figure 7.4 shows the route structure for DDSP as of mid-2011. Installations in blue were served by scheduled trucks, and those in red were not. The sizes of the circles indicate the relative volumes of the installations. Many of the routes go from DDSP to just one installation, although a few routes that go through the Georgia, Alabama, and Florida region stop at multiple installations.

Figure 7.5 demonstrates the powerful potential of linking non-served installations with each other and/or adding them as additional stops on existing routes. For example, a scheduled truck could stop at Charleston, Beaufort, Jacksonville, and Moody AFB, at a frequency of four times per week for the first three and stopping three times per week at Moody. Charleston and Beaufort would then go from not being serviced by scheduled trucks to four trucks per week,

Figure 7.3
The Cost and Performance Impact of Addressing Scheduled Truck Leakage—An Example from Ft. Bragg, CY 2010

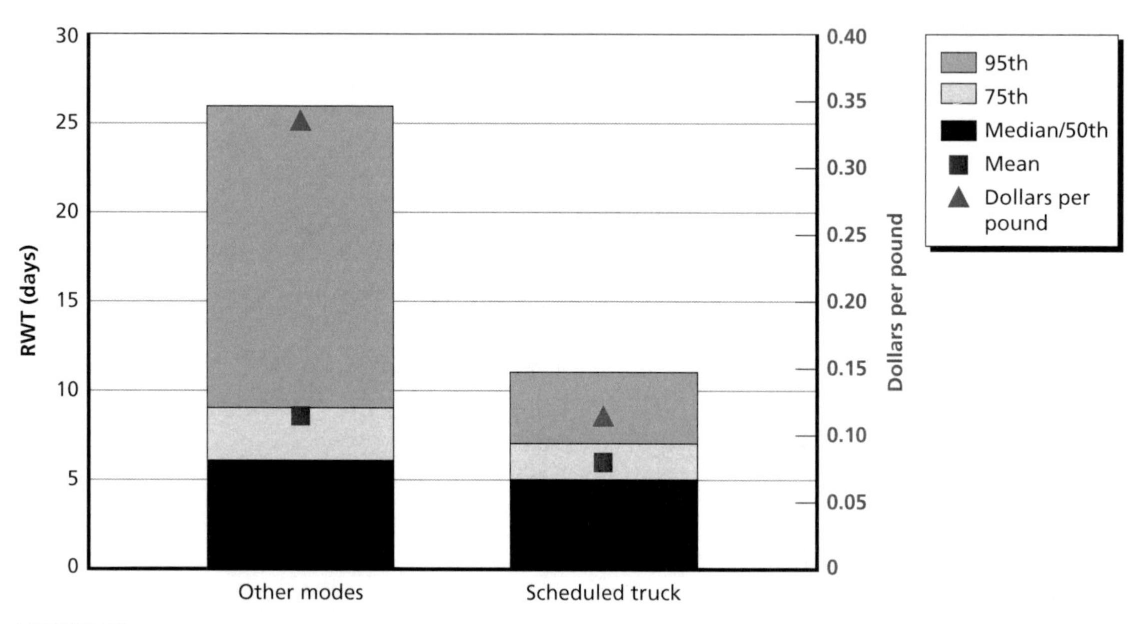

RAND *TR1274-7.3*

Figure 7.4
Mid-2011 Scheduled Truck Routes from DDSP

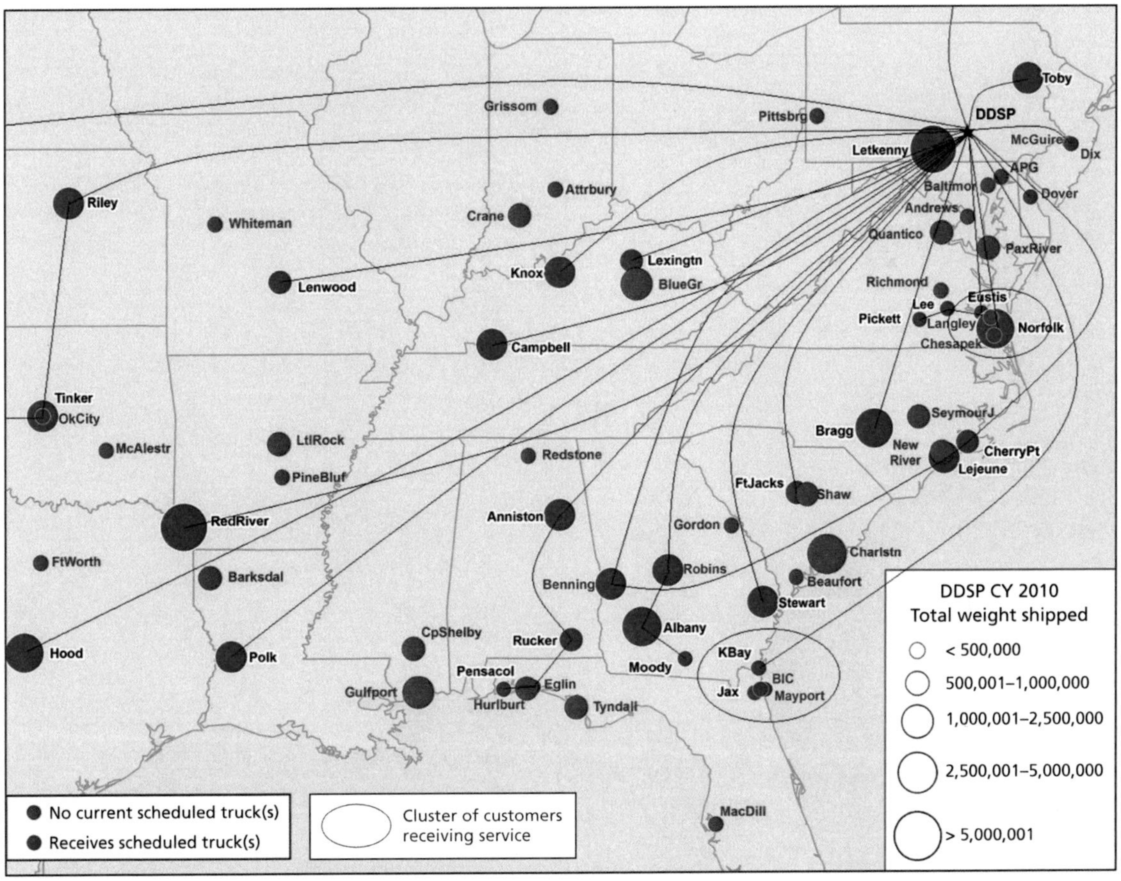

RAND *TR1274-7.4*

and the frequency of service to Jacksonville and Moody would increase by one truck per week to each. For these four installations, the average RWT would decline by 15 percent, with the aggregate annual cost of transportation to serve them dropping from $1.5 million to $0.4 million or 73 percent.

To fully examine the potential of optimizing routes for a given pattern of demands and stock positioning, RAND researchers, in the course of research for the Army, developed a route planning optimization software program. The input to the tool is installation and supply-activity-specific demand data by DC source. It then builds all possible scheduled truck routes that meet specified frequency, utilization, performance, and cost criteria based upon constraints and planning factors designed to ensure route feasibility. These constraints and planning factors include Federal Motor Carrier Safety Administration regulations (e.g., driving time per day), road networks, and factors such as typical travel speed and time per stop to drop off shipments. The tool then uses a greedy algorithm to select the best set of possible routes, focusing on minimizing total cost.[4] The output is the proposed route network, with each route having a defined set of installations, drops per installation, and frequency.

[4] As a greedy algorithm, the selected set may not be the truly best or optimal set but rather it should be a good set that is either optimal or close to being so, with relatively little difference in cost and performance from the truly best set.

Figure 7.5
An Example of an Improved Scheduled Truck Route with Multiple Installations

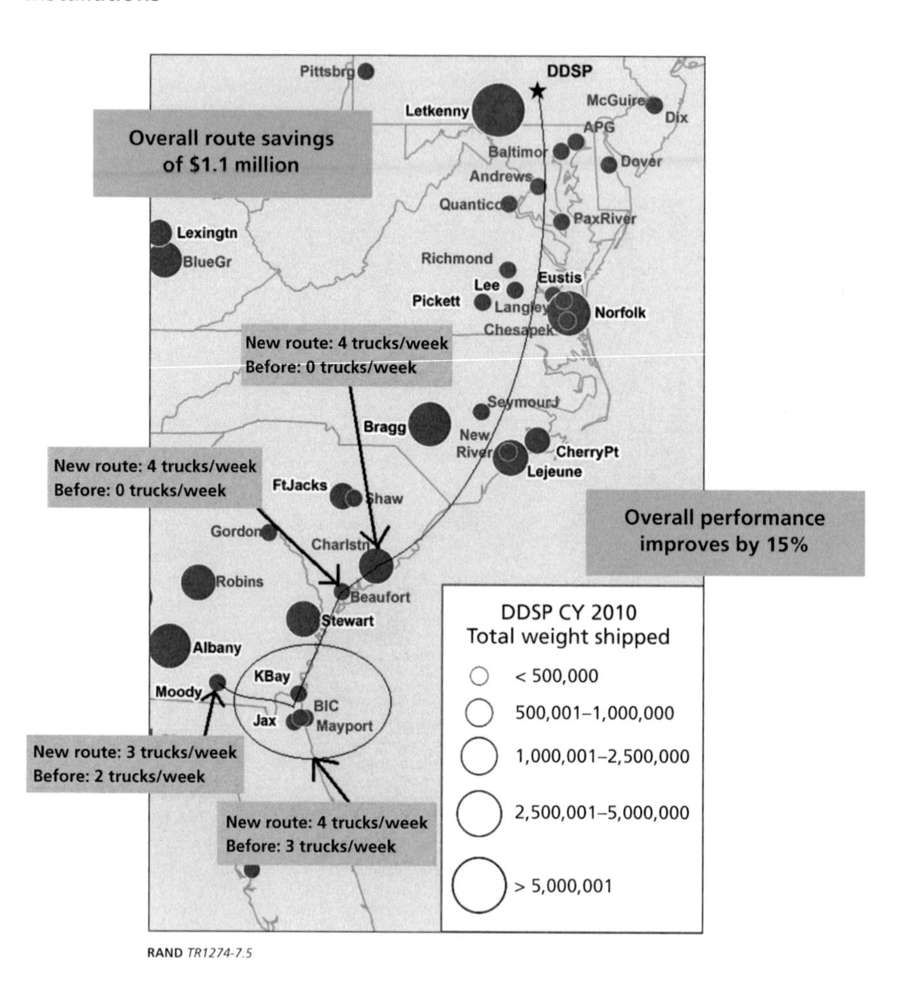

RAND *TR1274-7.5*

Figure 7.6 provides example output showing how the DDSP structure shown in Figure 7.4 would change to minimize cost subject to ensuring at least three trucks per week for each installation and no degradation in performance for any installation. Three trucks per week is the minimum frequency that enables meeting the IPG1 TDD goal for CONUS of seven days, but this constraint does not preclude the use of four or five trucks per week if needed to support the volume on a route. Most of the installations not served by scheduled trucks would be added to a scheduled truck route with this improved route structure, indicated by the lines or routes going through most of the red circles. Note that almost all routes would become multi-installation routes.

Implementing this structure in CY 2010 along with a similarly optimal structure for Defense Distribution Depot San Joaquin, CA (DDJC), was projected to save $10 million of the $52 million in transportation costs to serve CONUS active installations. Using CY 2011 data produced similar results. These route structures would have produced some improvement in RWT for low-priority shipments and would have held IPG1 RWT steady, primarily switching these latter shipments from air to scheduled truck.

Figure 7.6
An Example of an Optimized Scheduled Truck Route Structure from DDSP

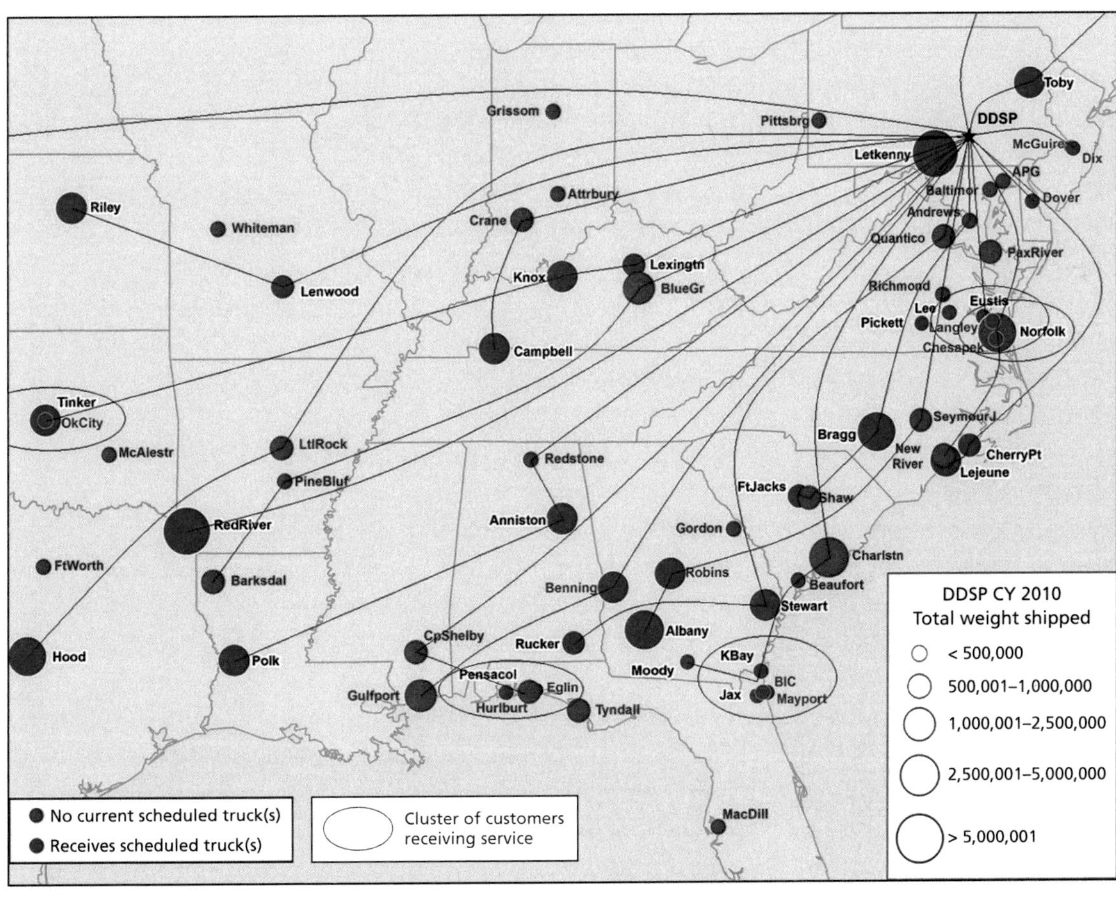

RAND *TR1274-7.6*

Applying this route optimization tool demonstrates how integrating DC consolidation planning with transportation planning and integrating transportation planning across installations that span services, taking a distribution system view, offers the potential for significant savings. If each supply activity selects the mode of service based upon its own needs and demands, then the overall system will be suboptimized from both efficiency and performance standpoints. Instead, to achieve the best total system solution requires changes in policy, procedures, oversight and diagnostic metrics, and the adoption of this type of route planning tool.

The recommended new policy is based upon a paradigm of the support provider choosing the best method of service that meets customer needs. In this case, customer needs are defined by TDD standards. Scheduled trucks become the best solution when a frequency that can meet the TDD standards allows sufficient utilization given a route's volume so that the average shipping cost per pound becomes less than it would be if priority-differentiated transportation modes were employed. In other words, scheduled truck utilization needs to be such that the cost per pound becomes lower than it would be were a mix of the other modes used. When combined, this leads to a policy that says:[5]

1. If the route volume supports well-utilized scheduled trucks that meet TDD standards, all customers on the route will have their shipments on the truck.
2. Otherwise, apply priority-differentiated transportation modes.

Customers would no longer have the option of opting out as long as the service provider meets TDD standards with scheduled trucks. It becomes up to the service provider to determine what truck utilization level is necessary for scheduled trucks to be the most efficient solution to achieve TDD standards. When route volume will not support efficient use of scheduled trucks at sufficient frequency, then the service provider should use the best shipping mode for each shipment based upon its priority, weight, size, special shipping considerations, and any opportunities for consolidation of shipments.

To implement such a policy, a central planning organization would need to be designated to determine the optimal route structure using a standard, automated route planning tool.[6] Determining the optimal route structure should be done on a periodic basis to capture shifts in demand, with changes also made to account for planned events such as unit deployments and stationing changes. Additionally, standard service provider and customer coordination processes are necessary to ensure route designs account for local needs and to ensure smooth, coordinated receipting operations.

Effective execution is necessary to ensure TDD standards are met and efficiency expectations are achieved. To ensure this, metrics for monitoring and control are needed. Performance metrics should be route delivery time for the supplier and transportation segments—divided into these two separate segments for diagnostics, truck utilization, and shipping cost per pound. Additional diagnostic metrics would include the percentage of shipments on the next possible scheduled truck and the percentage of shipments to installations on scheduled truck routes on scheduled trucks, which would help check for leakage.

---

[5] This policy recommendation has been incorporated into *DoD Supply Chain Materiel Management Procedures*, which calls for DLA to develop scheduled truck networks based upon the principles described here and to make their use standard practice, with exceptions only in accordance with policy guidelines (DoD Manual 4140.01, draft as of March 2012).

[6] In 2012, DLA distribution initiated a project with RAND to transfer the scheduled truck network planning code described in this chapter to DLA for use in a production environment.

CHAPTER EIGHT

Integrating Supplier and Transportation Management

Currently, for classes II, IV, and IX for items stocked in DLA DCs, for FDT or inbound freight, the transportation cost is included in the acquisition prices of the items. In conjunction, almost all such freight is contracted for as freight-on-board (FOB) destination, which means the supplier arranges and pays for the shipping to get the item to the DLA DC. As a result, there is no coordination across suppliers to consolidate shipments, even when the production sites are geographically proximate or along the same route to a DC. Inbound freight is not being managed from a total system integration standpoint. Additionally, with the use of FOB destination, suppliers cannot take advantage of DoD's transportation contracts, which would likely be valuable for smaller suppliers without the scale and resulting leverage to negotiate with transportation providers to achieve very low rates similar to those in DoD's contracts. Together these two issues present a potential opportunity for improved transportation, procurement, and supplier management integration for potentially lower total supply chain costs: DoD managed inbound freight, which could offer route consolidation and rate opportunities. This would apply the principles of leveraging scale advantages and aligning decisionmaking with the best systems view.

However, since inbound freight is included in acquisition prices, the transportation cost is not transparent; DoD does not know what its suppliers are paying for inbound freight. Therefore, estimating the financial benefit of a shift to DoD management of inbound freight cannot be done directly.

Instead, to estimate what the potential savings could be if inbound freight were consolidated across suppliers to enable more efficient transportation, we developed a database of one year of inbound shipments from suppliers to DCs. The objective was to determine the potential value of shifting LTL shipments to full truck load (FTL) shipments by using either static or dynamic "milk-run" routes that would consolidate shipments from multiple supplier locations bound for the same DLA DC. As Figures 8.1 and 8.2 show, there are significant supplier concentrations that deliver to the same DLA DC. Figures 8.1 and 8.2 show, respectively, the location of each supplier that shipped materiel to DDSP and to DDJC in 2010, with the sizes of the circles representing the weight shipped from each.

After producing a database of each shipment in terms of weight, we estimated the likely mode of each inbound shipment, split among FTL, LTL, and small package (assuming they are all ground shipped). Then we estimated the cost of the shipment based upon the origin and destination using DoD rates for the mode and distance. Then we parametrically examined the cost impact of shifting 25 percent to 100 percent of the LTLs to FTL in 25 percent increments. Below a given amount of weight, roughly estimated at 150 lb, the cost of an extra stop for a truck would not be worth the potential reduction in the transportation rate, so we did not

Figure 8.1
DDSP CY 2010 Supply Base

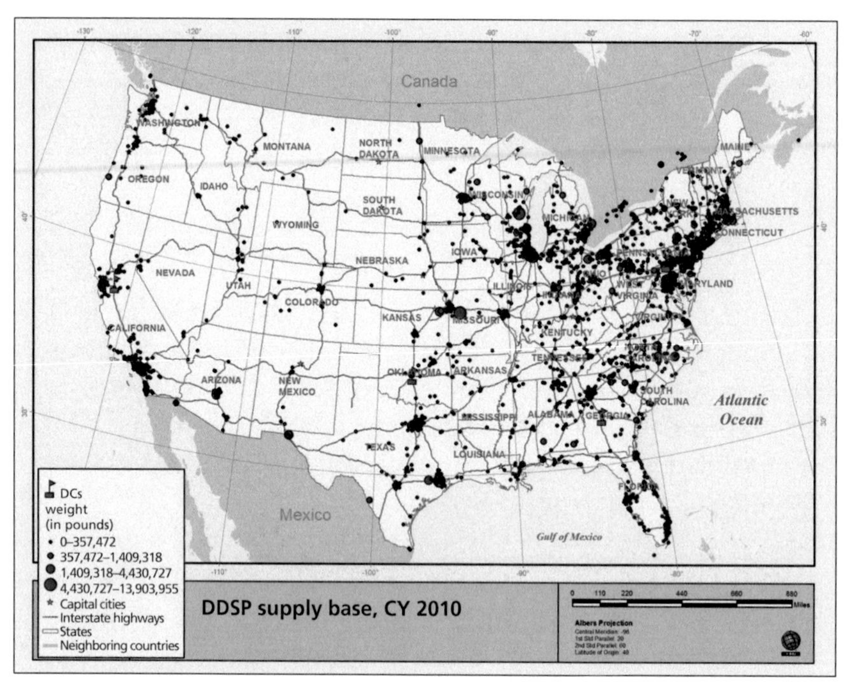

RAND *TR1274-8.1*

Figure 8.2
DDJC CY 2010 Supply Base

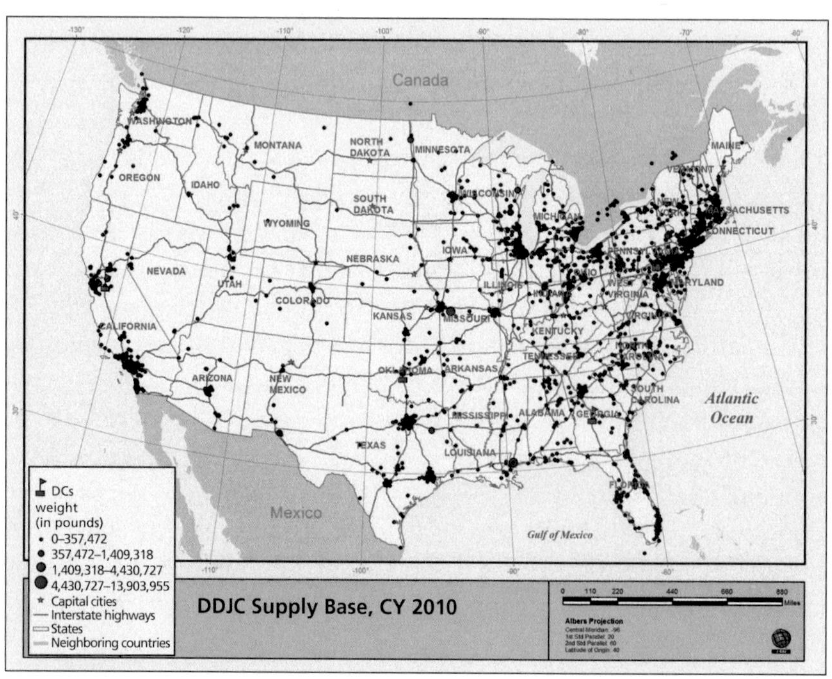

RAND *TR1274-8.2*

include the small package shipments. In this analysis, the baseline inbound freight costs came out to $43 million with savings ranging from $5 million to $20 million as 25 percent to 100 percent of LTL shipments are consolidated to FTL.

As an alternative, we also looked at what would happen if all shipments were sent initially to the closest SDP, without front-end consolidation, and then cross-docked to the intended buyback location(s). In effect, this would make one SDP the only buyback location for every NIIN, with other SDPs replenished via stock transport orders (STOs). This shortens the LTL leg, reducing these costs and allowing for most of the distance to be covered by FTL. Including cross-docking charges based upon DLA net landed costs, which are an activity-based set of costs it uses to estimate the costs of each DC transaction and the cost of storage, this would save about $9 million or roughly the same as shifting half of the LTL to FTL in the first estimate in the prior paragraph.

Additional savings would be possible if some DoD suppliers are paying higher shipping rates, in which case the baseline value would be higher. Also, if suppliers could use DoD shipping rates, some savings might accrue on the smaller replenishments shipped via small package carriers.

To achieve either form of savings, DoD would need the ability to shift to FOB origin, which would mean that DoD arranges for and pays the transportation. A regulatory review suggests some Federal Acquisition Regulation (FAR) hurdles, depending upon interpretation. Contracts below $150,000 and above this threshold need to be considered separately, because that is the simplified acquisition threshold below which there is a streamlined approach to contracting and contract management designed to reduce the administrative burden using different regulations. In doing so, though, the FAR calls for FOB destination contracts for these items. However, this could actually be contrary to the reduced burden purpose of the simplified acquisition procedures and other instructions that say, "Procurement below simplified threshold is to be inspected by the contractor." Above the threshold, the FAR allows for more discretion in contract terms, with contracting officers given discretion and instructions to evaluate contract terms on the basis of overall cost. Some regulations suggest use of the FOB term that would produce the lowest cost:

- Contract administration office (CAO) instructions must "result in the most efficient and economical use of transportation service and equipment."
- "The contracting officer shall determine FOB terms generally on the basis of overall cost . . . ."
- "The contracting officer shall consider the availability of lower freight rates to the Government for FOB origin acquisitions."

However, the regulations restrict this flexibility for some items that the FAR states must be FOB destination. These include perishable subsistence and medical supplies, commodity market items, certain lumber and steel shipments, and indefinite delivery type contracts. The FAR also specifies that when quality assurance (QA) is at the destination, acceptance should be too. Combined with the requirement for inspection at the destination for some items, this is generally consistent with the requirement for FOB destination.

However, given that some regulations call for finding the best overall value, which should include the possibility of government transportation contract rates and consolidation of transportation across suppliers, the regulations should be reviewed for clarification of interpretation, deconfliction of apparent opposing regulations and objectives, and potential change to allow for FOB origin when it would be the best value solution. This is common in the private sector, where liability issues are handled through contractual terms and allowances for expected rates of loss and damage.

CHAPTER NINE

Positioning Materiel Based on Total Costs

DLA employs a hub-and-spoke distribution network. In this network, the SDPs, which are intended to serve as CONUS regional hubs, are generally the destinations of inbound freight from suppliers or what are termed buyback locations. The SDPs directly support operational forces, such as tactical units, and replenish the spokes in most cases—FDPs in support of industrial activities and FDDs that provide the bigger, heavier, higher-demand items overseas. The major exception is when the FDP is the only major user of an item in a region. In this case, the FDP becomes a buyback location.

Stock Transport Orders

DLA moves materiel among DCs using STOs. There are two types of redistribution actions. One type of STO consists of moves to execute the hub-and-spoke system as designed, with replenishments of spokes from their supporting hubs; these are called planned deployments. DLA also uses what are called dynamic deployments, which are STOs that move stock that is excess in one location to another DC that needs it—from a planning standpoint. Local excess can develop when demand changes for a DC's designated customers, such as from restationing of units or shifts in the nature of operations. Dynamic deployments apply to all combinations of hubs and spokes except designated hub-spoke combinations, which would be replenished via the planned deployments. So materiel might be moved via dynamic deployments from one hub or SDP to another, from an SDP to a spoke outside of its region, or from a spoke to another spoke or an SDP. DLA has standard business rules in place to determine when to execute planned and dynamic deployments.

As part of the project described in Chapter One, opportunities were sought to identify options to cost-effectively improve stock positioning to further improve scheduled truck potential. In the course of doing this, it became clear that STOs were not being used as cost-effectively as possible from an integrated supply chain standpoint. For example, when looking at STOs between the two largest SDPs, it was found that the vast majority of STO quantities had quite low total weight—the product of the quantity and weight per item, resulting in spending more on materiel handling than the reduction in transportation cost from avoidance of out-of-area transportation that was being gained from these STOs. There were also a small number of potential STOs not executed that would have been cost-efficient or for which it would have been more cost-effective to buy more inventory at the shortage location.

Concurrently, DLA had also concluded that STOs were not being used cost-effectively, and a DLA team that one of the authors joined was formed in the fall of 2011 to examine how to improve STO cost-effectiveness. The team completed its work in December 2011, with recommendations quickly leading to changes in the DLA STO business logic implemented in February 2012.

Improving STO Business Logic

The resulting case study further demonstrates the need to consider tradeoffs in not only inventory, materiel handling, and transportation costs such as described earlier for FDD stockage, but also procurement workload, ensuring that decisions reflect total cost and are aided by decision support tools that take all costs into account. This is also a case in which functional areas of an organization were focused on their functional metrics without feedback on other costs and effects. While this chapter differs from the other efficiency opportunity chapters by describing an improvement already made, it is still included because the team directly applied the concepts discussed in this report and the conceptual total cost logic can be further extended to DLA stock positioning planning and to service stock positioning and redistribution planning, offering further opportunities.

For planned deployment STOs, in 2011, while the system computed optimal replenishment quantities (using EOQ-like logic) for shipments to the FDPs and FDDs, called coverage durations, if insufficient stock was available at the hub to fill this optimal quantity, any amount of stock down to a quantity of one could be shipped. For small items, this can increase costs by increasing the number of replenishment actions and thus issues and receipts in the SDPs, FDPs, and FDDs. Instead, the demanded materiel could be issued directly to the customer from its supporting SDP, avoiding these inefficient replenishment actions.

The logic for dynamic deployment STO determination had two flaws. First, there was a standard quantity that applied to all items above which a STO could occur. For smaller, lighter items, this quantity was too low to be economical, and for other, heavier items, it was too high, preventing economical STOs. STOs generate an additional set of issues and receipt touches and require transportation to redistribute the materiel. However, STOs enable the avoidance of transportation costs from a DC to customers out of its region that it does not generally support. For a dynamic STO to be cost-effective, one condition that needs to exist is that the STO costs should be lower than this alternative. Except for customers collocated with a DC, there is no difference in response time between the use of within-area and out-of-area transportation. Instead, different transportation options with different costs are used. For example, DDJC east of San Francisco supports Camp Pendleton, which is just north of San Diego. With Pendleton's high volume, scheduled trucks enable responsive, low-cost transportation from DDJC, averaging $0.13 per pound in FY 2011. If DDJC is out of stock for an item urgently needed at Camp Pendleton, it is most likely to come from DDSP, because the two DCs support similar types of customers. High-priority shipments will be sent using FedEx or commercial air freight, depending upon the item size and weight and the OQ, achieving the same wait time as getting it from DDJC. Yet it will cost much more—an average of $1.24 per pound in FY 2011. Even if the order is low priority, enabling ground shipment, the cost will still be higher than using the scheduled truck, with the overall average cost per pound of shipments from DCs other than DDJC averaging $0.88 in FY 2011. Given larger quantity STOs, opportunities to consolidate shipments, and the fact that STOs are not being shipped to serve immediate needs, relatively low transportation costs can often be achieved for them, particularly between SDPs and for designated hub-to-spoke combinations.

So the question becomes, do the net transportation savings from the STO outweigh the additional materiel handling cost? As the total weight (quantity multiplied by item weight) of a STO increases, the greater the net transportation savings become. For a very heavy item, a STO quantity of just one can produce net savings. For small, light items, the quantity needed to make up for the added materiel handling cost can be high. To illustrate this, here is an example. DLA approximates the costs of each issue and receipt transaction each year using activity-based costing. For a STO between 150 and 1,000 lb, the sum of the estimated issue and receipt costs would have been $119 in FY 2011.[1] If say, the transportation cost difference between out-of-area transportation on the one hand and STO transportation plus in-area transportation on the other hand were $0.30 per lb, which is roughly the system average, then it would take a STO quantity of 133 for a 3-lb item to break even.

The second problem with dynamic deployment was that when stock was considered excess but below the acceptable STO quantity, the inventory planning system would assume the STO would be executed anyway and account for it in supply and procurement planning as if it would occur. Thus, in determining the need for a procurement action at the shortage location, the system would count these assets even though they would not be moved. This would delay a procurement action to bring the location with a shortage to the planned stockage level.[2] The effect would be to force the customers of the out-of-stock location to receive materiel from a nonsupporting DC, which typically increases transportation cost. In some cases, this raises total costs when the more expensive out-of-area transportation outweighs the avoided additional inventory and workload buying costs for the shortage location. This introduces another condition in STO determination. Not only should a STO be more cost-effective than using out-of-area-transportation, it should also be more cost-effective than buying more materiel for the shortage location. Rather, the question should be which of three options would be most efficient: a STO, out-of-area transportation, or moving up a buy for the shortage location. The question should be broader than whether to do a STO when locally excess stock makes it an option; it should be which of these three options to pursue. For example, when a heavy item is inexpensive, buying more inventory for the shortage location will tend to be more cost-effective than doing a STO, because even the STO transportation cost becomes significant. When the heavier items are relatively expensive, a STO will tend to be much more cost-effective than buying more inventory to fill the local shortages or using out-of-area transportation. When items are cheap and light, buying more inventory will tend to be the lowest-cost option unless the potential STO quantity is quite large.

So there were two supply chain integration problems embedded in these business rules. The first, applying to both planned and dynamic deployments, was the lack of a method for considering the costs of all supply chain processes in decisions. Rather, the same rules in terms of quantities applied to all items and stock keeping units. Second, the business rules for dynamic deployments took this a step further and put precedence on avoiding procurement actions and inventory over avoiding out-of-area transportation without determining that doing so would always lead to the lowest total cost solution.

---

[1] Department of Defense, 2011.

[2] Note that this does not eliminate a procurement action; rather, it delays one, with the delay depending upon the relative size of the potential STO and the demand rate for the location.

In late 2011, the DLA team addressed both of these issues, without impacting its ability to meet customer wait time requirements. The team developed economically efficient STO business logic, implemented in early 2012. For dynamic deployments, this has two components. The first is determining whether a STO would be cost-effective to execute or not when there is local excess inventory. It compares the costs of three options and selects the lowest option from a total supply chain cost standpoint:

- The cost of a prospective STO, to include the costs of the STO transportation and materiel handling, to avoid an earlier procurement and out-of-area transportation.
- The cost of leaving the material at the excess location and using out-of-area transportation to fill customer orders from customers of the DC that has an inventory shortage, to avoid executing a STO or making an earlier procurement.
- The cost of making an earlier procurement to restock the location with a shortage, to include the increased procurement workload cost and increased inventory cost, and avoid a STO and out-of-area transportation.

Embedded in this logic is the notion of a minimum economic STO quantity. Below this quantity, which varies with the price of an item, its weight, and its demand level, it is more cost-effective to use one of the other two options. This concept also applies to planned deployment STOs—those used for intended replenishment from hubs to spokes. For planned deployments, there is still an economically optimal redistribution quantity—similar to an EOQ—that the system will plan to achieve. Below this there is now a minimum economical STO quantity.

There is one exception to the use of the minimum economic STO quantity for planned deployments. Some items at FDPs are stocked in a "retail" role for DS of depot maintenance operations. Since having the materiel at the supporting FDP instead of it having to come from the SDP following an order from maintenance reduces wait time, it is critical to ensure the retail items are on hand. So for these items, the minimum economic STO quantity does not apply.

Stockage Location and Replenishment Source Decisions

In the course of this analysis, the STO team also recognized that the same economically integrated considerations would apply in principle to determining whether to stock an item at a given location rather than using standard demand quantity and frequency-based criteria for all items, as has been the case. Trading off procurement, inventory, materiel handling, and transportation costs, one can determine whether to plan to stock an item at a given location or serve that location's customers from elsewhere using out-of-area transportation. In addition, one can determine whether to replenish that location directly from suppliers, making it a buy-back location, or via STOs from a hub or even another DC using these economic efficiency concepts. In February 2012, the STO team transitioned to develop the business logic to make such economic stock positioning and buyback location decisions. Additionally, this same total cost approach to stock positioning and repositioning decisions could be applied by the services.

CHAPTER TEN

Integrating Financial Policy with Network Design and Inventory Planning

All DoD field, retail, and tactical organizations generate serviceable returns through one of two general categories. The first is that sometimes end users order materiel they determine later that they do not need, which they return to their supporting tactical or retail supply activity (or sometimes they cancel the order too late, after it has been shipped from the wholesale provider). This could have been from a maintenance misdiagnosis, ordering some things in anticipation of a need that does not develop, mistakes, and the like. The second general category is inventory requirement changes at the tactical or retail level due to changes in operations, supported equipment, unit configuration, or other factors. This can result in some items in inventory no longer being needed. Thus, customer returns and local inventory changes both create local excess conditions at tactical or retail warehouses. While improving processes to reduce returns and inventory churn should always be pursued, some level will be a natural course of doing business. So then the question becomes, how can serviceable returns be managed most effectively? In the DoD case, financial management methods in combination with information integration shortfalls are inhibiting supply chain performance.

Up to a point, it makes sense to keep actively demanded local excess in place to be drawn down. Otherwise, it should be sent back to a central point for reuse. For DLA-managed items though, the services transfer money to DLA when they receive this materiel. They can get credit for returning it, but the average level of credit is low—or more accurately, credit is offered on a low percentage of potential returns. When a service has an item that is excess to its inventory needs, in accordance with service-specific business rules, it offers it for return to DLA. DLA then responds with an offer of credit equal to the latest acquisition cost of the item (or the standard price less the cost recovery rate—commonly called the *surcharge*), an offer to take back the item without credit (but paying the transportation charge), or refuses the return. If a service keeps an item, it can hold it or dispose of it. Credit is offered when the DLA inventory position is at or below the maximum inventory level for an item.[1] When DLA's inventory position is beyond a specified level, then the offer of return is refused. In the case when credit is given and the service has to order the same item again, the service in effect pays the surcharge a second time. So the magnitude of the surcharge could affect the service decision to return the item for credit, even when "full" credit is given, because for some items the surcharge can be substantial. If the service returns an item for no credit (or if it disposes of an item after a return denial) and then someone else in the service orders the same item later, the service has to pay the full price again. Instead, if the service keeps the item in its inventory accounts, it can reissue

[1] This is defined as the approved acquisition objective.

the item to the second customer without this second expenditure, saving service budget money. As a result, the services tend to keep much of this materiel when their forecasts suggest some potential for reuse. This is sometimes kept in tactical supply activities (e.g., when likely to be reissued in a short period of time), centrally on an installation, centrally in theater, or centrally for the service, often managing this materiel with retention limits.

The consequence of this practice, though, is some redundancy in distribution system capabilities and, perhaps more importantly, masking of demand for DLA planners. When the service has this retention stock, it generally uses it to satisfy demands, at least in the same region, before sending requisitions to DLA. Thus, for example, if a service generates a spike in excess for a given item, it may not order from DLA for a period of time, creating both forecast error for that period as well as potentially artificially suppressing future forecasts. Overall, the demand signal becomes more erratic, creating forecasting and planning difficultly.

There are two potential solution paths to this problem. The first is ensuring that information on service retention stocks of DLA-managed items are integrated into DLA planning systems. The data are available in service systems, along with underlying demand data. These systems are available to DLA. However, integrating the data for planning purposes is a manual process. This is something that could be automated, resolving the forecasting and planning problem. This would still leave shadow distribution capabilities. A path to resolve this portion has been explored by the Army and DLA. It would consist of returning the materiel to DLA DCs but keeping track of the number of assets that would still be in Army (service) ownership. In fact, a similar practice has been adopted in OCONUS FDDs. Currently, the Army still controls the issue of its assets of materiel from the DLA FDD. In the future, it is possible that DLA could issue the Army-owned inventory to fill demands from any service and integrate the assets into its supply planning.

The second path would be changing credit policy to remove the incentive to the services for keeping retention stock of DLA-managed items. Current transfer pricing signals are leading to inefficient behaviors rather than an efficient, integrated supply chain. Past research suggests that the low credit stems from undervaluing of serviceable assets and over-avoidance of one type of visible risk at the expense of creating other less transparent risk. DLA does not want to be left with excess inventory stemming directly from returns—the visible risk. In doing so, it creates inventory management risk stemming from the demand-masking problem. However, a partial barrier to this solution is that many DLA items are DVDs. Most could be stocked in CONUS DCs and then issued, with associated changes in the requisition sourcing logic to issue these prior to going to the supplier. However, BRAC 2005 required "disestablishing all other supply functions" besides contracting in the DLA supply centers along with "disestablishing storage and distribution functions for tires, packaged petroleum, oils, and lubricants, and compressed gases," so it is considered to be a requirement that these types of items would still have to stay under service ownership if returned to DLA DCs (as is being done now overseas) or kept in service-managed warehouses.[2]

Changes in the management of retention stock that would eliminate shadow distribution and warehouse capacity would continue a DoD trend of rationalizing DC capabilities and warehouse capacity. As with improved implementation of OCONUS FDD stock position-

[2] These items include tires; packaged POL, and compressed gases. Department of Defense, *Base Closure and Realignment Report, Volume I, Part 2 of 2: Detailed Recommendations*, May 2005. Defense Base Closure and Realignment Act of 1990, as amended (Part A of Title XXIX, Public Law 101-510; 10 U.S.C.).

ing, the attendant elimination of "lateral" shipments from DLA-managed item, service-owned retention inventory is a second significant source of potential savings identified in the Strategic Network Optimization initiative.

CHAPTER ELEVEN

Conclusions and Overall Recommendations

Fundamental to achieving supply chain integration and pursuing actions consistent with total supply chain optimization as opposed to process or functional optimization is always thinking about doing so, whether in management of the supply chain or its personnel, policy development, process design, and everyday decisionmaking. This starts with ensuring that workforce members understand how they affect the rest of the supply chain, receive feedback on their effects on other processes and their effects on the total supply chain, and have the tools to make integrated supply chain decisions. The framework laid out in this report in Chapter Four is intended to provide a basis for such understanding. This supply chain framework—describing an integrated structure; the key roles of each organization, process, and function within it; and how they depend upon each other—should be incorporated into DoD supply chain materiel management policy as a new DoD pamphlet. This framework creates the structure within which to work and the foundation upon which to build an integrated supply chain design, with enabling mechanisms holding it together in the way intended. The policy evaluation framework provided in this report in Chapter Three is an aid to be used to help ensure that the layers of policy fit together well and that the details meet the overall intent, following a consistent set of principles. Every person in DoD logistics and supply chain management can be an integrator; the challenge is broadly establishing a mindset that leads everyone in the workforce to focus first on the overall supply chain's performance—to be a systems thinker—instead of being confined within process, functional, or organizational walls.

There are several changes in emphasis that DoD should pursue to achieve supply chain integration and improved performance and efficiency. The first is increased attention to supplier lead times and order quantities and improving them through collaboration and integration with suppliers. An important step is being taken by including lead time in newly proposed OSD supply chain enterprise metrics. In conjunction, the role of procurement personnel in driving inventory must be recognized to a greater degree, giving them appropriate responsibility and accountability for the effects of contract terms and supplier performance and capabilities on inventory. In addition, OSD should launch a new initiative to determine how purchasing and supply management practices could be improved, building on DoD strategic sourcing and supplier relationship management initiatives, to achieve lead time and OQ reductions. However, DoD should remain cognizant of other costs as well to ensure that total costs go down. Related to this is ensuring a tight integration between demand, supply, and repair planning for reparable items to ensure the total supply of carcasses in the closed loop reparable system is kept to the minimum necessary to support readiness. Thus, a second initiative should examine reparable inventory management and the closed loop supply chain across the services to identify opportunities for improvement. In 2012, the ASD(L&MR) launched two studies to take on these issues with one focusing on improving consumable supply chain management in DLA and another focusing on reparable item management across the services.

The second change should be an increased focus on stock positioning as a key component of supply chain integration. This should include the incorporation of stock positioning in policy, the broad adoption of stock positioning metrics—to include addition in the OSD supply chain enterprise metrics, and constant consideration of all supply chain costs in making inventory planning and repositioning decisions.[1] The metrics should pertain to OCONUS FDD stockage, stockage at CONUS SDPs in support of scheduled truck customers, and FDP retail stockage in support of industrial activities. Improved stock positioning is at the heart of a number of important DoD supply chain initiatives, such as Strategic Network Optimization, DSO supply alignment, and the BRAC 2005–based transition to DLA ownership and management of retail stock in support of maintenance depots, along with having an important interplay and potential for leverage with a scheduled truck network improvement effort based upon Chapter Seven. Yet, despite the frequency with which stock positioning is the crux of improvement initiatives, emphasis remains limited, as reflected in metrics and by the lack of goals for stock positioning.[2]

Building off of an increased emphasis on stock positioning should also be an increased emphasis on how distribution planning affects transportation utilization, with feedback given to distribution planners and even procurement personnel. The former influences utilization through shipment consolidation plans with respect to customer groupings and the latter can influence it through contracts that require providers to use the Defense Transportation System. As part of this, shipment consolidation criteria with regard to when to use scheduled trucks, pure or mixed pallets, and pure or mixed containers should be added to supply chain materiel management policy.[3]

Related to all of these is ensuring organizations have the breadth of budgets that give them the degree of freedom to pursue the course of action that will optimize the supply chain and are correspondingly responsible for budgets that they drive the consumption of. For example, this might mean shifting OCONUS second destination transportation budgets to the various supply management organizations given that stock positioning tends to drive the use of airlift versus sealift for delivery overseas—either to an FDD or the customer. This would enable them to trade off inventory investment dollars with transportation dollars, choosing the solution that requires the lowest overall budget. A review of supply chain organizational budget categories and the effects that each organization has on costs should be conducted to determine where there is misalignment, with changes made accordingly. If multiple organizations have significant effects on a cost category, then the organization that has the greatest ability to affect the cost should have the budget authority, with the other organization(s) having cost metrics reflecting their effects, giving them visibility and responsibility for how they are affecting the budgets of other organizations. Aligning budget authority and organizational effects should

---

[1] As described earlier in different places in this report, the stock positioning recommendations in this report have been incorporated into DoD Manual 4140.01 (draft as of March 2012), and DLA has incorporated total supply chain cost considerations into OCONUS stock positioning planning and stock repositioning logic and is in the process of revising its CONUS stock positioning logic accordingly.

[2] During the course of this study, DLA developed OCONUS stock positioning metrics and goals.

[3] Policy to develop these criteria has been incorporated into DoD Manual 4140.01 (draft as of March 2012).

also be part of the design process when standing up new organizations or changing organizational designs.

Finally, continued progress toward end-to-end information sharing, to include outside the boundaries of DoD is crucial to ensuring integration, improving performance, and avoiding perturbations in integration, as shown in the second case study in Chapter Two. This includes ensuring each organization knows what information it produces—and more importantly, could produce that it is not—that would be valuable to its upstream and downstream partners. It also includes ensuring that organizations develop capabilities to utilize this information to the full potential.

Beyond these policy changes, calls for shifts in management emphasis, and related changes in enabling mechanisms, particularly metrics, there are some specific actions that DoD could pursue or further examine for potential supply chain efficiencies based upon integration. The first is implementing a new process and planning tool for optimizing the CONUS scheduled truck network in conjunction with policy for its standard use when cost-effective. The second would be more detailed examination of the potential benefit of DoD-managed FDT along with review of whether regulations should be changed to better support FOB origin. The third should be continuing efforts to incorporate total cost considerations into inventory positioning decisions, within DLA and the services. The fourth should be determining how to best eliminate inefficiencies from service retention stock of DLA-managed items.[4]

DoD directly controls and operates much of its supply chain. For the relevant parts of the system, designing the right structure, having the right policy, and aligning metrics, feedback, and accountability with the overall system's performance are crucial. Externally, contractual incentives can also be applied, with it being critical that they be designed to encourage actions that optimize the supply chain rather than narrow considerations. Whenever gaps occur in the form of clear examples of process or functional optimization at the expense of decreasing the efficiency and effectiveness of the overall supply chain, these various layers—structure, policy, and enabling mechanisms—should be explored to find the problem and any misalignments.

---

[4] As of the writing of this report, efforts to act on all of these recommendations except reviewing regulations affecting the use of FOB origin are under way.

APPENDIX A

Legal and Regulatory Environment for FDT Alternatives[1]

Introduction

This appendix explores the level of discretion DoD contracting officers have in various situations concerning the choice of transportation methods for FDT. In particular, it focuses on whether and under what circumstances DoD contracting officers may solicit and award contracts on an FOB origin basis, as opposed to the routine FOB destination contracts that are now the norm for classes II, IIIP, IV, and IX. As a threshold matter, the ability to solicit and award FOB origin contracts is necessary if DoD-managed FDT were to be pursued.

Three main sections of this appendix trace the three legal-regulatory sources of contracting governance. The first is a brief review of statutory authorities contained in the United States Code. The second is a more detailed discussion of the contract management requirements contained in the FAR. This discussion delves into both the contract requirements and the inspection and acceptance requirements that accompany the government's QA role. The third section discusses how the FAR applies to DoD activities and supplements the regulatory scheme outlined in the FAR with the Defense Federal Acquisition Regulation Supplement (DFARS), DTR, and also the DLA Troop Support Guiding Principles for Acquisition (DGPA).

Where appropriate, the appendix highlights areas of vagueness and possible legal confusion that might cast doubt on whether a contracting official is able to choose alternative contract terms for first destination transport. For the purposes of this discussion, the scope of the appendix is limited to the contracting requirements associated with first destination shipments that are both originating from and destined for locations within CONUS.

Statutory Sources Governing Federal Contract Administration

The federal government's procurement actions are, in the broadest sense, governed by the acquisition statutes. Although the annual authorization and spending legislation tackles the yearly budgetary issues for the federal government, they also are the common mechanism for establishing more permanent federal procurement policies. However, although they are the source of the government's contracting authority, they do not contain the details of the contracting procedures affecting procurement. Those are a product of regulation (discussed in the next sections).

[1] We thank our supporting author Geoffrey McGovern for the text in this appendix.

The amalgamated federal procurement statutes are codified in Title 41 of the United States Code (U.S.C.). Title 41 is a piecemeal collection of permanent laws governing public contracts. In addition to Title 41, legislation specific to the armed services' procurement actions are contained in Title 10, section 2301, et seq. The following is a brief commentary on both titles' contents as they apply to contracting.

Title 41 applies to all federal agencies. The title contains the organic law authorizing the Office of Federal Procurement Policy (OFPP), which coordinates government procurement, and outlines the OFPP administrator's authorities (§1121). Much of the title pertains to broadly applicable standards and guidance: cost accounting practices, public-private competitions, recordkeeping requirements, and contract dispute mechanisms. Major legislative revisions, such as the increase in the simplified acquisition threshold to $100,000 (passed in the Federal Acquisition Streamlining Act of 1996; currently set at $150,000), are permanently codified in Title 41.

As is common with legislation governing the federal agencies, much of the day-to-day activities associated with contracting and specific procurement policies are delegated to the agency heads. Specifically, acquisition regulation has been delegated by statute to the administrator of General Services, the Secretary of Defense, and the administrator of National Aeronautics and Space. These individuals jointly maintain the FAR (discussed in detail, below).[2]

Title 10 applies only to DoD agencies and governs the armed services. Section 2311 permits the head of the each Title 10 agency (DoD, the Department of the Navy, the Department of the Army, the Department of the Air Force, the Department of Homeland Security, and the National Aeronautics and Space Administration) to delegate the agencies' procurement functions to appropriate officers. This has been delegated at the DoD level to the Defense Procurement and Acquisition Policy (DPAP) office for all acquisition and procurement policy matters. DPAP manages DFARS (discussed below).

For purposes of this appendix, transportation-related procurement policies are generally absent from the statutes, other than to specify that the government can enter into special rate agreements (government rate tenders) with commercial carriers. Whereas procurement quantities do receive attention (41 U.S.C. §3310), this is likely a reflection of political interest in procurement. Transportation likely has not been the focus of similar scrutiny. The only relevant sources of transportation-related procurement policies are found in 49 U.S.C. 10721 (concerning the government ability to ship freight at no cost or reduced rates on railroads), and in 49 U.S.C. 13712 (concerning the government ability to ship freight at no cost or reduced rates on a motor carrier, water carrier, or freight forwarders[3]). Hence, the applicable statutory authority concerning transportation as a matter of federal procurement has been delegated and exclusively is contained in federal regulation.

[2] For an example of this type of delegation, see 41 U.S.C. §3906: Cost-reimbursement contracts, stating that "the Federal Acquisition Regulation shall address the use of cost-reimbursement contracts." This level of abstraction is typical for the statutory language when there is a clearly identified authority (such as OFPP) that is responsible for the policy minutia.

[3] A *freight forwarder* is defined in 49 U.S.C. 13102 as

> a person holding itself out to the general public (other than as a pipeline, rail, motor, or water carrier) to provide transportation of property for compensation and in the ordinary course of its business—(A) assembles and consolidates, or provides for assembling and consolidating, shipments and performs or provides for break-bulk and distribution operations of the shipments; (B) assumes responsibility for the transportation from the place of receipt to the place of destination; and (C) uses for any part of the transportation a carrier subject to jurisdiction under this subtitle. The term does not include a person using transportation of an air carrier.

Federal Acquisition Regulation

Section 1303 of 41 U.S.C. created OFPP for the federal government and authorizes the administrator of General Services, the Secretary of Defense, and the administrator of National Aeronautics and Space to "issue and maintain federal acquisition regulation" (known as the FAR). The FAR codifies federal policies for acquisition of supplies and services by executive agencies (and is contained in Chapter 1 of Title 47 of the Code of Federal Regulations). The FAR is the product of the federal rulemaking process and is subject to revision pursuant to the Administrative Procedures Act (60 Stat. 237).

FAR Simplified Acquisition Procedures (Part 13)

Transportation-related issues in the FAR are dispersed throughout the regulation. The first mention of transportation costs in the FAR occurs in Part 13. Part 13 describes the government's policy for simplified acquisitions procedures. Simplified acquisition procedures are used for the acquisition of goods and services valued below $150,000. Meant to remove much of the red tape that can significantly delay government procurement of essential and routine materiel, these simplified procedures specify general principles for contract solicitation, evaluation, and award/documentation. Moreover, acquisitions valued below the threshold are set aside for small business concerns (FAR 13.003(b)(1)).

A review of the regulation indicates that both the desire for a speedy process and the small business set-aside may play a role in the government's ability or inability to realize savings through DoD-managed FDT. The qualification is based on two possible ways to interpret the regulation. The reference to transportation costs in Part 13 concerns evaluation of contractor quotations or offers. For precision, the language is reprinted here:

> 13.106-2 Evaluation of quotations or offers.
>
> (a) General.
>
> (1) The contracting officer shall evaluate quotations or offers—
>
> (i) In an impartial manner; and
>
> (ii) Inclusive of transportation charges from the shipping point of the supplier to the delivery destination.

One reasonable reading of this regulation is to conclude that the offer itself must indicate a *price* that is inclusive of transportation charges. This would bundle the materiel cost and the transportation cost. In the quotation submitted to the government for below-threshold orders, the contractors' cost would then include, *but not necessarily itemize*, the cost of FDT. This has the potential to obscure the actual cost of the materiel.

The competing interpretation would conclude that the *evaluation process* must consider the transportation charges. Such an interpretation would place the onus on the contracting officer to identify transportation costs and to take those costs into consideration when evaluating offers. This would obviate the need for the contractor to obtain transport for the goods and allow the government to use favorable, pre-negotiated rates with a preferred carrier. However, counter to the time-saving intention of the simplified procedures, this interpretation would require contracting officers to *investigate* the transportation costs associated with each offer, rather than merely evaluate competing offers.

The tension between these two interpretations comes from the increased time and process required to manage transportation for many small contracts and the intended simplification of procedures for low dollar value procurements. However, because acquisitions below the threshold level are set aside for small businesses, supplanting the contractor obtained transportation rates for government-negotiated rates may offer attractive savings.

Whether Part 13 of the FAR holds the potential for substantial savings will depend, in part, upon how many of the vendor contracts are low-value procurements subject to the simplified procedures. It may also depend upon whether contracting officers can successfully solicit low-value offers that break out transportation costs (since this is not, at the moment, required). Even if such a breakout of costs were required, it remains to be seen whether the contractors are willing to accept contract awards that do not include transportation to destination, or whether they launch a protest claiming that the FAR's simplified procedures preclude such a searching review of transport costs.

While recognizing the possible interpretation that the simplified procedures require a price that includes transportation costs, the remainder of the FAR regulations suggests that the alternate interpretation is the most consistent. That is, the requirement in Part 13 mandates that cost be included in the evaluation process, but not necessarily as part of the contractor offer. For example, in Part 13.302-1, when describing the procedures for purchase orders under the simplified acquisition procedures, the FAR states that purchase orders "shall specify f.o.b. destination for supplies to be delivered within the [continental] United States *unless there are valid reasons to the contrary*" [italics added for emphasis]. This is the clearest statement in Part 13 that even the rapid, red-tape-free procedures are intended to allow contracting officers discretion. Cost is a valid reason and is indicated to be especially important in Part 47.

FAR Transportation Regulations (Part 47)

Given that contracting officers are allowed discretion over choice of FOB terms, the question remains as to what information might inform this discretionary judgment, and whether there are any limits to the exercise of discretion. Parts 46 and 47 of the FAR establish regulations for QA and transportation, respectively, for procurement of supplies, both of which are grouped under a subchapter heading "Contract Management."

FAR transportation policy in Part 47 states that the CAO is required to "ensure that instructions to contractors result in the most efficient and economical use of transportation services and equipment" (47.101(b)).[4] This suggests that cost factors are regularly to be included in the awarding and administration of contracts, including the determination of FOB terms. This is a general statement of principle, supported by the more direct regulation in 47.304-1(a) that "*The contracting officer shall determine f.o.b. terms generally on the basis of overall costs . . .* " [italics added for emphasis].

It is important to note that the terms of the contract are established in the solicitation of contractor offers. The choice of FOB terms must be stated in the *solicitation* (which may avoid the vagueness in interpretation of the simplified acquisition procedures above). To do so, the

---

[4] Specifics of how the CAO will manage this responsibility are delegated to the agencies. The FAR notes that more specific policies for DoD are provided in U.S. Transportation Command, *Defense Transportation Regulation—Part II: Cargo Movement*, DTR 4500.9-R, May 13, 2011a (which is discussed in the next section of this report).

contracting officer must prepare in advance an assessment of whether FOB origin or FOB destination terms are more advantageous to the government (a term that includes consideration of overall cost, as well as traffic management concerns).[5] In doing so, "the contracting officer shall consider the availability of lower freight rates [Government rate tenders] to the Government for FOB origin acquisitions" (47.304-1(c)).

However, while the regulations just described seem clear, there is confusion in the FAR as to whether contracting officials actually can specify the FOB terms in the solicitations. While 47.304-1(b) states, "Solicitations shall specify whether offerors must submit offers FOB origin, FOB destination, or both; or whether offerors may choose the basis on which they make an offer," subsequent language in 47.304-2(a) states, "Solicitations shall provide that offers may be submitted on the basis of either or both FOB origin and FOB destination and that they will be evaluated on the basis of the lowest overall cost to the Government." This latter regulation could be read to mean that solicitations *must* allow the contractor to submit offers on any FOB terms they choose. The second regulation applies only to shipments within CONUS. Still, there is uncertainty whether the contracting officer can solicit only FOB origin contracts or whether the choice of terms must be left up to the contractor.

A reasonable (if debatable) interpretation that takes the entire Part 47 into account would conclude that the contracting officer has the discretion to solicit offers on the terms that are most advantageous to the government. If an FOB origin contract were the most advantageous, it would not be reasonable to solicit offers that allow the contractor to dictate FOB destination terms.

Apart from solicitation issues, when the government procures goods and does not require the contractor to bear the cost of transportation, then the bill of lading[6] is the required shipping method for domestic shipments (47.101(a)). This means that common carriers are the preferred service providers for government-ordered shipments. Per 47.101(h): "when a contract specifies delivery of supplies FOB origin with transportation costs to be paid by the Government, the contractor shall make shipments on bills of lading . . . either at the direction of or furnished by the CAO or the appropriate agency transportation office." Presumably, this coordination is done to obtain the full benefit of the government tender rate.

In fact, FAR policy expressly contemplates the possibility that an FOB origin contract can be combined with transportation priced via the government rate tender. 47.104-2 allows fixed price contracts with FOB origin terms to require contractors to *prepay* transport charges via a commercial bill of lading at the government tender rate. The contractor will then be reimbursed "for the direct and actual transportation cost as a separate item in the invoice." These rates are not allowable for fixed-price, FOB destination contracts. However, the government tender rate may be applied for shipments made under cost reimbursement contracts (47.104-3).

There are some limitations on the contracting officer's discretion to choose delivery terms:

- When destinations are tentative or unknown, the solicitation shall be FOB origin only (47.304-1(d)).

---

[5] Oddly, a subsequent section of the FAR [47-305-2(a)] requires the solicitation to specify that offers "will be *evaluated* on the basis of the lowest overall cost to the Government." Traffic management concerns are not part of the offer evaluation process.

[6] As used in the FAR (47.001), a *bill of lading* is "a transportation document, used as a receipt of goods, as documentary evidence of title, for clearing customs, and generally used as a contract of carriage."

- When the size or quantity of supplies with confidential or higher security classification requires commercial transportation services, the contracting officer shall generally specify FOB origin acquisitions (47.304-1(e)).
- If acceptance is at the destination, then the terms must be FOB destination (47.304-1(f)).

The last restriction, concerning acceptance of goods, is a matter of QA. QA matters are addressed in Part 46 of the FAR and are reviewed briefly below.

FAR Quality Assurance Regulations (Part 46)

QA procedures are designed to ensure that the products and services the government acquires meet the contract terms and standards. General QA processes involve the inspection of the goods, the acceptance of the goods from the contractor or carrier, and warranties about the goods' quality. For definitional purposes, acceptance means the taking of ownership of materiel by a government official or authorized representative.

Under FAR regulations, three main policies are essential: First, all contracts must provide for inspection. Second, all inspection must be conducted by or under the direction of government personnel. Third, QA (meaning inspection) is almost always conducted before acceptance of the goods (46.102). QA terms are established in the solicitation and agreed to in the contract itself.

The nature of the QA activities depends on the nature of the goods being acquired. For example, goods acquired at or below the simple acquisition threshold ($150,000) are to be inspected by the contractor, unless the government has a need to test the supplies, to judge the adequacy of the contractor's internal procedures (46.202-2), or to inspect the goods upon reaching the destination for kind, quantity, damage, etc. (46.404). For the procurement of goods above the simple acquisition threshold, when those goods are not especially complex or critical items, the FAR prescribes contract terms that specify the QA process. A special category of higher-level contract quality requirements applies to especially complex or critical items.[7]

As they relate to transportation policies and the choice of destination terms, the FAR's QA regulations matter to the extent they establish the place of acceptance. Recall that if acceptance of the materiel must be at destination, then the contract terms must be FOB destination (47-304-1(f)). This requires an understanding of when acceptance must take place at destination.

The FAR generally assumes that acceptance and QA (inspection) are normally done at the same time and in the same place; however, this place is a matter of contractual terms. FAR 46.503 states that

> each contract shall specify the place of acceptance. Contracts that provide for Government contract quality assurance at source shall *ordinarily* provide for acceptance at source. Contracts that provide for Government contract quality assurance at destination shall *ordinarily* provide for acceptance at destination [italics added for emphasis].

Hence, under ordinary circumstances (but see below), this regulation makes the place of acceptance contingent upon the place of QA.

---

[7] Per the FAR (46.202-4), "Examples of higher-level quality standards are ISO 9001, 9002, or 9003; ANSI/ISO/ASQ Q9001-2000; ANSI/ASQC Q9001, Q9002, or Q9003; QS-9000; AS-9000; ANSI/ASQC E4; and ANSI/ASME NQA-1."

In limited cases, some aspects of QA (inspection) must take place at destination (which could require FOB destination contracts). For these cases,

> Inspection shall be performed at destination under the following circumstances:
>
> - Supplies are purchased off-the-shelf and require no technical inspection;
> - Necessary testing equipment is located only at destination;
> - Perishable subsistence supplies purchased within the United States, except that those supplies destined for overseas shipment will normally be inspected for condition and quantity at points of embarkation;
> - Brand name products purchased for authorized resale through commissaries or similar facilities (however, supplies destined for direct overseas shipment may be accepted by the contracting officer or an authorized representative on the basis of a tally sheet evidencing receipt of shipment signed by the port transportation officer or other designated official at the transshipment point);
> - The products being purchased are processed under direct control of the National Institutes of Health or the Food and Drug Administration of the Department of Health and Human Services;
> - The contract is for services performed at destination; or
> - It is determined for other reasons to be in the Government's interest (46.403).
> - Unless a special situation exists [exceptionally complex or critical materials], the Government shall inspect contracts *at or below the simplified acquisition threshold* at destination and only for type and kind; quantity; damage; operability (if readily determinable); and preservation, packaging, packing, and marking, if applicable (46.404-3(b)(1)) [italics added for emphasis].

The interaction of these rules produces some curious results. If, for example, the government is procuring commercially available goods or goods valued at less than $150,000 (by law, from a small business), then inspection must be limited to destination (per 46.403 or 46.404-3(b)(1)), and *ordinarily* shipping terms must be FOB destination (per 47-304-1(f)). Assuming that this is an ordinary circumstance, this regulation holds even if the government would otherwise benefit from an FOB origin contract by having the contractor prepaying the shipping at the government tender rate. The solicitation terms for such a contract, looking backward through the process, must solicit only an FOB destination contract despite the potential government savings from alternative terms.

The use of the term "ordinarily," however, suggests that an alternative reading would allow the contracting official to procure goods through FOB origin terms if doing so would be to the government's advantage. The availability of the government tender rate for lower shipping costs might reasonably be used to justify extraordinary contract terms that allow for off-the-shelf goods to be shipped FOB origin (even though they will be inspected at destination). This is, however, speculative. The FAR is not clear on whether cost savings from alternative delivery terms is sufficient justification for a deviation from the FOB destination requirement when acceptance must be at destination.

For these cases, it is also possible to break with the standard assumption that inspection and acceptance take place at the same time and place. Although inspection for the items listed above must be done at destination, some of these inspections are duplicative. For example, the true quality inspection for acquisitions under the simple acquisition threshold is performed at the source by the contractor. Government inspection at destination is of a lesser nature, to ensure quantities and look for damage in transit. For these circumstances, and consistent with FAR 46.501, acceptance could take place before delivery, following the contractor inspection.

Similarly, some purchases require QA at the source (46.402). But because there are no transportation regulations that require FOB origin contracts when acceptance occurs at the source, we do not get the same curious outcomes as above.

Department of Defense Regulations

There is general acceptance in DoD regulations that FAR contracts are appropriate for DoD shipments. DTR Chapter 201(M.3) explicitly notes, "A FAR contract is suitable for any DoD traffic regardless of commodity or transportation requirement. They are best where there is a requirement for recurring traffic or a long period, a large volume, or an oversized movement." Apart from this, DTR contains procedures for issuing commercial and government bills of lading once the government is procuring transportation from a carrier.

In addition to DTR, FAR 1.301 authorizes agency heads to issue regulations implementing and supplementing the FAR. For DoD, the major supplement is DFARS (although DLA Troop Support has issued additional supplements; see below). DFARS does not substantially alter the transportation regulations about the choice of FOB terms contained in the FAR. But tangential QA regulations in DFARS do affect shipping choices, and the effect may be at odds with other FAR and DFARS requirements about cost-effectiveness.

For example, DFARS Part 246.402 specifies that contracts or delivery orders for less than $300,000 should *not* require government contract QA at the source (and 246.404 explicitly applies this to acquisitions at or below the simplified acquisition threshold). (This rule is waived if DoD regulation specifically requires QA at the source, if there is a memorandum of agreement about QA at the source between the Defense Contract Management Agency [DCMA] and the agency requesting the procurement, or if the contracting officer determines that QA must be at the source for other critical reasons.) For the ordinary case, QA inspection for contracts valued less than $300,000 will take place at destination.

Normally, this requirement would mean that both QA inspection *and acceptance* are to be held at destination (because FAR 46.503 ordinarily requires acceptance at destination if QA is at destination). Moreover, if acceptance is at destination, then the FOB terms must be FOB destination (FAR 47.304-1(f)). However, this could be in conflict with general policy goals because requiring FOB destination may increase the overall cost to the government. If so, there would be a direct conflict between the QA inspection terms in DFARS 246.4, the FOB destination requirement of FAR 47.304-1(f), and other transportation regulations, including

- FAR 47.101(b), which requires contracting instructions to "result in the most efficient and *economical* use of transportation services and equipment"
- DFARS' own policy statement that the defense agencies shall "develop and manage a systematic, *cost-effective* Government contract quality assurance program" [italics added for emphasis] (DFARS 246.102(1))
- FAR 47.304-1(a), which states that the "contracting officer shall determine FOB terms generally *on the basis of overall cost*" [italics added for emphasis].

At the moment, there is no clear instruction in DFARS or the FAR that overrides the sweeping language of DFARS Part 246.402 requiring QA inspection at destination for contracts under $300,000 (other than the special cases noted above). It is unclear whether the substantial cost savings that may be gained from alternative FOB transportation terms would override 246.402.

To complicate matters further, however, DLA Troop Support's DGPA (discussed further, below), which governs acquisitions made by DLA Troop Support and their field operations, recognized that contracts valued less than $300,000 are to undergo QA at destination *but* allows contracting officials to change the inspection and acceptance terms to require inspection and acceptance at the source in instances of the following:

- A sole source or best value offer being contingent on source I/A [inspection/acceptance] and/or F.O.B origin when destination I/A is sufficient, but efforts to negotiate I/A at destination are unsuccessful;
- destinations that are unknown;
- combining DVD CLIN(s) [contract line item number] with stock CLIN(s) on the same award;
- when potential awardee is listed on the DCRL [Defense Contractor Review List] as requiring source I/A; or
- a memorandum of agreement with the DCMA.

This gives DLA Troop Support the ability to change the inspection and acceptance terms to inspection and/or acceptance source, thereby possibly avoiding the FAR requirement for FOB destination terms. But this is contingent upon one of the five instances, above. The memorandum of agreement with DCMA might be the most interesting option to pursue.

DLA Troop Support Guiding Principles for Acquisition

Finally, as mentioned above, DLA Troop Support issues its own regulations about transportation, which generally track the FAR.[8] DGPA Part 47.301-2 specifies that the transportation officer is supposed to recommend the FOB basis on which to solicit offers. 47.304-2 specifies, for shipments for purchases within the Simplified Acquisition Threshold within CONUS, "the contracting officer shall use good business judgment when determining the requested FOB terms." For acquisitions above the threshold, "Quotations/offers shall be solicited on the basis of either or both FOB origin and FOB destination delivery terms."[9] DGPA cites FAR 47.304.2 as the basis of this rule (which might help to resolve interpretation issues discussed above).

[8] The authority to issue the DGPA is stated in the intro to the guide (DSCP DGPA):

> The Defense Supply Center Philadelphia (DSCP) Guiding Principles for Acquisition (DGPA) is issued by the direction of the Commander, DSCP pursuant to the authority contained in Subpart 1.3 of the Federal Acquisition Regulation (FAR), Defense FAR Supplement (DFARS) and the Defense Logistics Acquisition Directive (DLAD). The DGPA implements/supplements the FAR, DFARS, DLAD and other DoD acquisition publications. It establishes policies and delegation of authority governing the acquisition of supplies and services by DSCP and its field organizations.

[9] Furthermore, the DGPA seems to further endorse the possibility of FOB origin terms for acquisitions under the threshold in 47.306-1P(a), which governs procedures for transportation officials obtaining transportation cost information when evaluating FOB origin offers for values below the simplified threshold.

The DLA DGPA goes into more detail, specifying that perishable subsistence and medical supplies subject to in-transit deterioration shall be FOB destination only. Likewise, for subsistence commodity market items, FOB destination terms are required. Lumber and steel shipments may be required to use FOB destination terms, depending on whether the shipments are outside CONUS. And, finally, indefinite delivery type contracts should be FOB destination only (Subpart 47.304-2).

Conclusion

There is enough tension between competing regulations to significantly blur the requirements for contracting FOB terms. While the general principles at work seem to require attention to cost as one of the factors, if not the predominant factor, when considering FOB terms, other regulations seem to restrict a contracting officer's ability to select FOB origin terms for reasons other than cost. These restrictions are mainly a product of the simplified acquisition threshold, which in an effort to cut through the red tape of acquisition regulations, may have made it more difficult for contracting officials to procure materiel at the lowest cost.

Going forward, it will be important to solicit feedback from the contracting, transportation, and general counsel communities at DoD, as it is likely that all three groups are operating under different interpretations of the regulatory environment.

APPENDIX B

Inventory Performance Analysis

Generalized Boosted Model

This analysis examines the effects of different factors on measures of inventory performance: material availability/backorder rate, inventory turns, maximum months of supply on hand, and maximum length of stock-out periods. Linear regression will not work well since the data do not fit the assumptions (e.g., linearity and normality), so another statistical technique, the Generalized Boosted Model (GBM), that does not face the same problems was employed.

The results that follow indicate the relative influence of the factors considered. The relative influence of a factor or covariate is the percentage reduction in absolute error attributable to that factor.

The population of items covered in the analysis consists of the top 20,000 NIINs (DVDs are excluded, based upon a July 2011 snapshot of what items are DVD versus DLA direct) in terms of the extended value of demand in CY 2010 that had four or more demands resulting in 16,469 items. The analysis is based on demands, backorders, and inventory levels for these NIINs between August 2008 and July 2011.

GBM Description

GBM adds together many simple functions of the covariates (usually piecewise constants or simple regression trees) to estimate a smooth function of a large number of covariates. GBM is an algorithm that at each iteration adds to the previous iteration model a simple regression tree model. At every step, the algorithm essentially looks for a small adjustment to the prior iteration model that improves the fit of the model to the data.

The number of iterations determines the model complexity, and it is derived from the data. Generally the best number of iterations is determined by stopping rules that choose the number of iterations that maximizes the predictive performance of the model on an independent data set. This analysis was run using a five-fold cross-validation. See Tables B.1 and B.2.

Because the final GBM model is a sum of regression trees, it has many of the good properties of regression trees. Trees can handle continuous, nominal, ordinal, and missing covariates. They are flexible because they can capture nonlinear effects and interaction terms. Trees are invariant to one-to-one transformation of the covariates and can handle large number of covariates even if correlated among each other.

Table B.1
Relative Effects by Factor (out of 100) in DLA NIIN Statistical Analysis

Factor	Average	Backorder Rate	Inventory Turns	Longest Stock-Out Period (Weeks)	Maximum Peak (Months of Supply)
Unit price	26.5	24.0	31.4	21.7	25.9
Demand variability	20.9	16.0	11.7	17.2	16.4
Lead time	17.6	20.8	10.4	21.2	17.5
Demand level	15.9	20.2	24.3	21.3	20.5
Supply chain	11.3	9.4	7.5	8.1	9.1
Forecast model	4.2	4.6	6.5	5.7	5.2
RMC	1.8	2.4	0.6	1.9	1.6
SPR item	0.6	1.2	6.8	1.2	2.4
Dominant customer	0.5	0.7	0.4	1.1	0.7
Provisioning item	0.4	0.3	0.2	0.5	0.4
Collaborative item	0.3	0.4	0.3	0.2	0.3

Table B.2
Range of Effects by Factor in DLA NIIN Statistical Analysis

Factor	Backorder Rate	Inventory Turns	Longest Stock-Out Period (Weeks)	Maximum Peak (Months of Supply)
Unit price	0.22	3.0	18	31
Demand variability	0.25	2.8	10	37
Lead time	0.26	1.4	16	27
Demand level	0.27	5.0	50	78
Supply chain	0.30 (C&T, sub)	1.0 (4 w/sub)	9 (15 w/sub)	10
Forecast model	0.05	0.5	11	5
RMC	0.01	0.0	0	0
SPR item	0.02	0.2	0	1
Dominant customer	0.01	0.1	0	1
Provisioning item	0.01	0.1	0	1
Collaborative item	0.01	0.1	0	1

NOTES: C&T is clothing and textiles, and sub is subsistence. For inventory turns and the longest stock-out period, the range of effects results for the supply chain factor is shown with and without the small number of stocked subsistence supply chain items included in the dataset.

References

Assistant Secretary of Defense for Logistics and Materiel Readiness, *Comprehensive Inventory Management Improvement Plan*, October 2010. As of February 23, 2012:
http://www2.dla.mil/j-6/dlmso/archives/jpiwg/meetings/28Sep10/DoD_ComprehensiveIM_ImprovementPlan071610.pdf

———, *DoD Supply Chain Materiel Management Procedures*, DoD Manual 4140.01, Volumes 1 through 11, draft as of March 2012.

Branson, Richard W., "High Velocity Maintenance Air Force Organic PDM: Assessing Backshop Priorities and Support," *Air Force Journal of Logistics*, Volume XXXIV, Numbers 3 and 4, June 2011, pp. 16-25. As of February 23, 2012:
http://www.aflma.hq.af.mil/shared/media/document/AFD-110608-004.pdf

Dail, Lieutenant General Robert T., General Benjamin S. Griffin, and General Norton A. Schwartz, "Transforming Priority Requisitions to Optimize Distribution," memorandum, October 12, 2006.

Defense Federal Acquisition Regulation Supplement, Code of Federal Regulations, Title 48, Parts 200–299. As of February 22, 2012:
http://ecfr.gpoaccess.gov/cgi/t/text/text-idx?sid=f3e034e4b164d3572508c9fecc4d4488&c=ecfr&tpl=/ecfrbrowse/Title48/48cfrv3_02.tpl

Defense Logistics Agency, "Hub & Spoke Operational Business Rules," memorandum, July 21, 2006.

———, "Agency Performance Review," February 2011 and September 2011 .

———, "Inventory Stratification Report," September 2011.

Defense Supply Center Philadelphia (DSCP), Guiding Principles for Acquisition (DGPA). As of February 20, 2012:
http://www.troopsupport.dla.mil/contract/dgpa/dgpatoc.asp

Department of Defense, *Defense Working Capital Fund, Defense-Wide Fiscal Year (FY) 2012 Budget Estimates Operating and Capital Budgets*, February 2011.

———, "Inventory Stratification Report," September 2009.

———, *Base Closure and Realignment Report, Volume I, Part 2 of 2: Detailed Recommendations*, May 2005.

Department of the Air Force, *United States Air Force Working Capital Fund (Appropriation: 4930), Fiscal Year (FY) 2012 Budget Estimates*, February 2011. As of February 22, 2012:
http://www.saffm.hq.af.mil/shared/media/document/AFD-110209-040.pdf

Department of the Army, *Army Working Capital Fund Fiscal Year (FY) 2012 President's Budget*, February 2011. As of February 21, 2012:
http://asafm.army.mil/Documents/OfficeDocuments/Budget/BudgetMaterials/FY12//awcf.pdf

———, "Army Metrics Submission," June 2011.

———, "Depot Maintenance Initiatives," 2011 Army Posture Statement, July 11, 2011.

Department of the Navy, *Fiscal Year (FY) 2012 Budget Estimates: Justification of Estimates Navy Working Capital Fund*, February 2011. As of February 23, 2012:
http://www.finance.hq.navy.mil/FMB/12pres/NWCF_BOOK.pdf

Deputy Secretary of Defense, "Supply Chain Materiel Management Policy," DoD Directive 4140.1, April 22, 2004. As of February 23, 2012:
http://biotech.law.lsu.edu/blaw/dodd/corres/pdf2/d41401p.pdf

Deputy Under Secretary of Defense for Logistics and Materiel Readiness, "DoD Supply Chain Materiel Management Regulation," DoD 4140.1-R, May 23, 2003. As of April 16, 2012:
http://www.dtic.mil/whs/directives/corres/pdf/414001r.pdf

DoD 4140.1-R—*See* Deputy Under Secretary of Defense for Logistics and Materiel Readiness.

DoD Directive 4140.1—*See* Deputy Secretary of Defense.

DoD Instruction 4140.01—*See* Under Secretary of Defense for Acquisition Technology and Logistics.

DoD Manual 4140.01—*See* Assistant Secretary of Defense Logistics and Materiel Readiness.

Dumond, John, Marygail K. Brauner, Rick Eden, John R. Folkeson, Kenneth J. Girardini, Donna J. Keyser, Eric Peltz, Ellen M. Pint, and Mark Y. D. Wang, *Velocity Management: The Business Paradigm That Has Transformed U.S. Army Logistics*, Santa Monica, Calif.: RAND Corporation, MR-1108-A, 2001. As of February 20, 2012:
http://www.rand.org/pubs/monograph_reports/MR1108.html

England, Gordon, "Redesignation of the Commander, United States Transportation Command as the Distribution Process Owner (DPO)," memorandum, May 8, 2006. As of February 23, 2012:
https://acc.dau.mil/adl/en-US/32312/file/6091/DPO%20Redes%20Memo.pdf

Federal Acquisition Regulation (FAR), Code of Federal Regulations, Title 48, Parts 1-51. As of February 21, 2012:
http://www.access.gpo.gov/nara/cfr/waisidx_07/48cfrv1_07.html

Fisher, Marshall, "What Is the Right Supply Chain for Your Product?" *Harvard Business Review*, March–April 1997. As of February 23, 2012:
http://www.computingscience.nl/docs/vakken/scm/Fisher.pdf

Folkeson, John R., and Marygail K. Brauner, *Improving the Army's Management of Reparable Spare Parts*, Santa Monica, Calif.: RAND Corporation, MG-205-A, 2005. As of February 20, 2012:
http://www.rand.org/pubs/monographs/MG205.html

GAO—*See* U.S. Government Accountability Office.

Held, Thomas, Lisa Colabella, Matthew Lewis, John Halliday, and Christopher McLaren, "An Assessment of Opportunities for Improving the Education and Career Development of Army Supply Chain Managers," unpublished RAND Corporation research, 2007.

Kaminski, Paul G., "Lean Logistics: Better, Faster, Cheaper," speech, Leesburg, Va., October 24, 1996. As of January 2, 2012:
http://www.defense.gov/speeches/speech.aspx?speechid=873

Office of the Deputy Assistant Secretary of Defense for Supply Chain Integration Supply Chain Metrics Group, "Proposed Enterprise Metrics as of 20 Dec 11," December 20, 2011.

Office of Management and Budget, *Guidelines and Discount Rates for Benefit-Cost Analysis of Federal Programs*, Circular No. A-94, October 29, 1992 (Appendix C, Revised December 2011).

———, "Table of Past Years Discount Rates from Appendix C of OMB Circular No. A-94," November 16, 2011.

Peltz, Eric, and Aimee Bower, "The Drivers of Operational Readiness Rates: A National Training Center Analysis," unpublished RAND Corporation research, 2001.

Peltz, Eric, Kenneth J. Girardini, Marc Robbins, and Patricia Boren, *Effectively Sustaining Forces Overseas While Minimizing Supply Chain Costs: Targeted Theater Inventory*, Santa Monica, Calif.: RAND Corporation, DB-524-A/DLA, 2008. As of February 20, 2012:
http://www.rand.org/pubs/documented_briefings/DB524.html

Peltz, Eric, John Halliday, Marc Robbins, and Kenneth J. Girardini, *Sustainment of Army Forces in Operation Iraqi Freedom: Battlefield Logistics and Effects on Operations*, Santa Monica, Calif.: RAND Corporation, MG-344-A, 2005. As of June 7, 2012:
http://www.rand.org/pubs/monographs/MG344.html

Peltz, Eric, and Thomas Held, "Improving Readiness for Problem Ground Fleets," unpublished RAND Corporation research, 2003.

Peltz, Eric, and Marc Robbins, *Leveraging Complementary Distribution Channels for an Effective, Efficient Global Supply Chain,* Santa Monica, Calif.: RAND Corporation, DB-515-A, 2007. As of June 7, 2012:
http://www.rand.org/pubs/documented_briefings/DB515.html

Peltz, Eric, Marc Robbins, Kenneth J. Girardini, Rick Eden, John Halliday, and Jeffrey Angers, *Sustainment of Army Forces in Operation Iraqi Freedom: Major Findings and Recommendations*, Santa Monica, Calif.: RAND Corporation, MG-342-A, 2005. As of February 20, 2012:
http://www.rand.org/pubs/monographs/MG342.html

PRTM, *DOD Joint Supply Chain Architecture Annotated Briefing of Results and Repeatable Approach Release 2.0*, October 15, 2008.

Richey, R. Glenn, Jr., Anthony S. Roath, Judith M. Whipple, and Stanley E. Fawcett, "Exploring a Governance Theory of Supply Chain Management: Barriers and Facilitators to Integration," *Journal of Business Logistics*, Vol. 31, No. 1, 2010, pp. 237–256.

Robbins, Marc, Patricia Boren, and Kristin J. Leuschner, *The Strategic Distribution System in Support of Operation Enduring Freedom*, Santa Monica, Calif.: RAND Corporation, DB-428-USTC/DLA, 2004. As of February 20, 2012:
http://www.rand.org/pubs/documented_briefings/DB428.html

Under Secretary of Defense for Acquisition Technology and Logistics, *DoD Supply Chain Materiel Management Policy*, DoD Instruction 4140.01, December 14, 2011.

United States Code, Title 10, Armed Forces, August 10, 1956.

United States Code, Title 41, Public Contracts, January 4, 2011.

United States Code, Title 49, Transportation, October 17, 1978.

United States Government Accountability Office (GAO), Defense Inventory, *Opportunities Exist to Improve the Management of DOD's Acquisition Lead Times for Spare Parts*, Washington D.C., GAO-07-281, March 2007. As of February 23, 2012:
http://www.gao.gov/new.items/d07281.pdf

———, *Overview of Department of Defense's (DOD's) Comprehensive Inventory Management Plan DOD's Inventory Management Improvement Plan*, Washington, D.C., GAO-11-240R, January 7, 2011.

U.S. Air Force Global Logistics Support Center, "AFGLSC Monthly Performance Review," May 26, 2011.

U.S. Marine Corps Logistics Command, "CSCMP Marine Corp," May 2011.

U.S. Naval Supply Systems Command, "NAVSUP Business Metrics Review Session Logistics Support," December 22, 2010.

U.S. Transportation Command, *Distribution Process Owner Strategic Opportunities (DSO) Submission for: Supply Chain Operational Excellence*, 2009. As of January 2, 2012:
http://supply-chain.org/f/sce-awards/2009-US-Transcom-Submission.pdf

———, *Defense Transportation Regulation—Part II: Cargo Movement*, DTR 4500.9-R, May 13, 2011a.

———, USTRANSCOM Point Paper "Distribution Process Owner Strategic Opportunities (DSO)—Strategic Air Optimization (SAO) Opportunities," June 2011b.

———, USTRANSCOM Point Paper "Distribution Process Owner Strategic Opportunities (DSO)—Strategic Surface Optimization (SSO) Opportunities," June 2011c.

Wang, Mark Y. D., *Accelerated Logistics: Streamlining the Army's Supply Chain*, Santa Monica, Calif.: RAND Corporation, MR-1140-A, 2000. As of February 23, 2012:
http://www.rand.org/pubs/monograph_reports/MR1140.html

Wang, Mark, Jason Eng, Rachel Rue, and Jeffrey Tew, "Adapting Secondary Item Planning to Pull Production," unpublished RAND Corporation research, 2009.